MW00427602

Essays on the History of
Organic Chemistry

Essays on the History of
Organic Chemistry

Edited by JAMES G. TRAYNHAM

Louisiana State University Press

Baton Rouge and London

Copyright © 1987 by Louisiana State University Press
All rights reserved
Manufactured in the United States of America

Designer: Albert Crochet
Typeface: Linotron Aster
Typesetter: G & S Typesetters, Inc.
Printer: Thomson-Shore, Inc.
Binder: John H. Dekker & Sons

10   9   8   7   6   5   4   3   2   1

LIBRARY OF CONGRESS CATALOGING IN PUBLICATION DATA
Essays on the history of organic chemistry.

   Includes index.
   1. Chemistry, Organic—History.   I. Traynham,
James G.
QD248.E87   1986   547'.009   86-14377
ISBN 0-8071-1293-3

Portions of Chapters 2 and 3 have appeared, in somewhat different form,
in "Kekulé's Dreams: Fact or Fiction?" *Chemistry in Britain*, 20 (1984),
720–723, and "Biomolecular Handedness," *Chemistry in Britain*, 21
(1985), 538–545, respectively, and are reprinted by permission. Material
in Chapter 6 has appeared, in different form, as "The Physical Organic
Community in the United States, 1925–50," *Journal of Chemical Education*, 62 (1985), 753–757, and is reprinted by permission. The general
editor also gratefully acknowledges permission to reprint stanzas from
"Talking Redox Blues," by Howard M. Shapiro, Milan Bier, and C. F.
Zukoski.

Contents

Preface and Acknowledgments   vii

ALAN J. ROCKE
Convention Versus Ontology in Nineteenth-Century
   Organic Chemistry   1

JOHN H. WOTIZ and SUSANNA RUDOFSKY
The Unknown Kekulé   21

STEPHEN F. MASON
From Molecular Morphology to Universal Dissymmetry   35

O. BERTRAND RAMSAY
The Early History and Development of Conformational Analysis   54

JOHN A. HEITMANN
A New Science and a New Profession: Sugar Chemistry
   in Louisiana, 1885–1895   78

LEON GORTLER
The Development of a Scientific Community: Physical Organic
   Chemistry in the United States, 1925–1950   95

JAMES G. TRAYNHAM
The Familiar and the Systematic: A Century of Contention
   in Organic Chemical Nomenclature   114

JACK H. STOCKER
Chemage: A Compendium of Chemical Trivia   127

Notes on the Contributors   137

Index   141

Preface and Acknowledgments

In 1967 the Department of Chemistry at Louisiana State University organized and presented the first LSU Mardi Gras Symposium in Organic Chemistry. Held nearly every year since that time, the symposium is a full day of lectures and discussion; participants then have the opportunity to observe some of the revelry on the following day, Mardi Gras. During the 1970s, sponsorship of a few of the annual symposia was shared with Tulane University, the University of New Orleans, and Loyola University. These LSU Mardi Gras symposia in organic chemistry have apparently stimulated the organization of other annual symposia, also scheduled around Mardi Gras, in fields including theoretical chemistry, psychology, and theatre.

The symposia have been superb supplements to the regular course work and seminar programs at LSU and nearby universities and industries. The speakers, who come from many different universities and industrial laboratories to report on new developments in their research, have made substantial contributions in their areas, which include polymer chemistry, catalysis, new syntheses and synthetic methods, insect chemistry, and organic theory. In 1984 the focus of the symposium shifted from the present to the past. A one-day session on the history of organic chemistry could only be highly selective, but it did succeed in acknowledging the continuity of the discipline and in giving fresh perspective on the present. The papers presented, on topics in the nineteenth and twentieth centuries, provided carefully researched information and insights that were new to most chemists. This volume is the outgrowth of those papers.

Comments written over a decade ago by Aaron J. Ihde, who

was instrumental in the growth of the University of Wisconsin Department of Chemistry to national stature, are a particularly appropriate introduction to the essays herein.

> I would argue very strongly that it is time to take a positive approach toward history of chemistry. It is a subject which is important in its own right and needs no defense or apologies. . . . I am interested, and I believe most of us are, in the *education* rather than the *training* of chemists. The person who is merely trained to carry out analyses or syntheses can do his job quite satisfactorily without knowing any history of chemistry. On the other hand, the chemist who is in a position where he has significant responsibility for the planning of investigations needs to know something about the past history of chemical investigation and the development of chemical thought. Without such knowledge, he is merely a chemical technologist. ("Let's Teach History of Chemistry to Chemists!" *Journal of Chemical Education*, 48 [1971], 686–87)

Research in the history of chemistry may lead to controversy, just as research in chemical behavior does. The controversy in both cases may be of two kinds: disagreements on interpretation of the same data and disagreements on the data themselves. To different groups of chemists, each kind of controversy seems the more or the less resistant to resolution. Conclusions advocated in some of these chapters are contrary to longheld views, and some of them have recently stimulated controversy. The history of organic chemistry itself provides assurance that such disagreements usually lead to earlier and more careful reexamination of data and interpretation than would have occurred in their absence. We hope that the essays in this volume will illustrate the liveliness of research in the history of organic chemistry and will stimulate readers to reexamine some of the pivotal points in that history.

In recent years, Louisiana State University's financial support for the symposium has been substantially supplemented by gifts from chemical, petroleum, and publishing companies. The expenses of the sixteenth symposium, on which this volume is based, were paid with funds from three LSU units—the Department of Chemistry, the College of Basic Sciences, and the Center for Energy Studies—and from the following donors: Nalco Chemical Company; John Wiley and Sons, Inc., Publishers; Dow Chemical USA (Louisiana Division, R and D); Ethyl Corporation; Shell Chemical Company; American Chemical Society (Baton Rouge section); Ciba-Geigy Corporation (St. Gabriel Plant); Pennzoil Products Company; Allied Chemical; Enraf Nonius Service Corporation; Freeport Chemical Company; Merck Sharp and Dohme; Stauffer Chemical Company; American Cyanamid Company; and Georgia-Pacific Corporation. I gratefully acknowledge that support as well as an additional generous contribution by Ciba-Geigy Corporation (St. Gabriel Plant) to support the publication of this volume.

Ms. Juanita Miller and Ms. Stacy Cutrer patiently and cheerfully typed and retyped the manuscript. They eased my editing graciously, and I am grateful to them. Most of the drawings, and photographs of them for reproduction, were prepared expeditiously by Ms. Tina Roller of the Louisiana State University College of Basic Sciences Drafting Shop. I am also thankful for the instructive assistance of Ms. Catherine Landry, LSU Press editor, who tempered firmness with occasional acquiescence as she guided this book toward publication.

Essays on the History of
Organic Chemistry

ALAN J. ROCKE

Convention Versus Ontology in Nineteenth-Century Organic Chemistry

THE philosopher Maurice Mandelbaum once wrote: "To trace the essential phenomenological and inferential steps by means of which . . . our modern conception of the world has come to be what it is, remains one of the most comprehensive and challenging tasks for historians of science. . . . Such a task . . . [is] not only of historical but of epistemological significance."[1] It is such inferential steps that best illustrate the power of science—power not in a technological but rather in an intellectual sense. The facts of a science are interesting enough, but what one *does* with the facts is the real point, not to mention the beauty, of science. It is ineffably wondrous to trace the shapes of antediluvian rivers and Neolithic constellations, and to describe the social and sexual characters of Pliocene hominids, the drift of continents, the color and charm of quarks, the evolution of stars and galaxies, and the synthesis of dodecahedrane.

What, indeed, does a scientist do with the facts? The stock answer is that he categorizes and explains them by constructing scientific laws and theories. But these are much misunderstood entities, scientific laws being commonly regarded as nothing more nor less than theories that have somehow been "proven" by something called "the scientific method." No scientist has ever enunciated a law that is both universal and exact, though I do confess my personal belief in the imperishability, at least in overall outlines, of certain scientific theories such as evolution, heliocentrism, atomic theory, and relativity theory.

But the purpose of laws and theories is not so much to establish eternal verities as simply to explain. The nineteenth-century organic chemist Jean-Baptiste Dumas once wrote: "In chemistry, our theories are crutches; to show that they are valid, they must be used to walk. . . . A theory established with the help of twenty facts must explain thirty, and lead to the discovery of ten more."[2] In brief compass Dumas thus described three functions, or tests of the power, of a scientific theory: first, explanation of the facts used to set up the theory; second, "retrodiction," or reverse prediction, of previously discovered data not used in the initial creation of the theory; and third, the heuristic function of predicting unexpected phenomena as consequences of the theory.

Mandelbaum used the term *transdiction* to describe a fourth function of a certain class of theories: deduction of the existence of unobservable phenomena from observable ones. Transdictive inference is far more common than might be supposed. For instance, Alexandre Koyré pointed out forty-five years ago that pure inertial motion has never been seen—indeed in a strict sense it occurs only in an observerless thought experiment—yet the law of inertia has formed the fruitful basis for all systems of mechanics since René Descartes.[3] But for no science is transdiction more fundamental than for chemistry. The deduction of detailed molecular architectures from purely macroscopic manipulations of substances is among the most awesome accomplishments of modern science, indeed, of the human mind.

Thus far I have been writing from the perspective of ontological realism, the view that molecules really exist, independent of any observer, with specific structures that are accessible in some way to the tools, senses, and minds of scientists. I suspect that virtually all contemporary chemists operate on a basis of intuitive realism. However, many physicists and philosophers have serious doubts about the metaphysical basis of realist assumptions, and it is quite possible, and a century ago was even common, to do good chemistry from other than a realist standpoint. The nineteenth-century French chemist Pierre Duhem suggested that the goal of realist theory was to explain, and he defined *explanation* as the process of stripping reality "of the appearances covering it like a veil, in order to see the bare reality itself." He criticized this endeavor as blatantly metaphysical and asserted that physical theory should be nothing more than "an abstract system whose aim is to *summarize* and *classify logically* a group of experimental laws, without claiming to explain those laws." For Duhem and other representatives of the positivist tradition, theories are conventions rather than ontological explanations. They function mnemonically by ordering information logically and economically in our minds, they provide illuminating analogies between different branches of science and realms of nature, and they furnish a natural classification of data and laws. A natural classification is as simple as possible and reveals the highest number of analogies; predictions based on the system are validated by discoveries that fit logically into the taxonomy.[4]

I wish to propose a "natural classification" of organic-chemical theorists in the nineteenth century. In broad outline, during most of the century a desultory debate took place between conventional and ontological theorists, that is, between those who regarded the goal of theory as taxonomic convention and those who sought to approach the ultimate reality behind sensible appearances. But such a simple dichotomy will not do justice to the complexity of history, and at least one more taxonomic bifurcation is required. I shall define *orthodox* conventionalism as the viewpoint that it was impossible, in principle, to demonstrate progress in science in an ontological sense. A much larger class of organic-chemical theorists, however, represented a position I shall name *methodological* conventionalism. These scientists regarded the pursuit of conventional theory as the proper scientific method but entertained hopes for successful ontological theories. One way to view this distinction is to think of the orthodox camp as those who were intellectually committed to an operationalist, descriptivist, sensationalist, idealist, and positivist philosophy, and the methodological camp as those who took the conventionalist route as a mere *pis aller*, an expedient in an imperfect world.[5]

The ontological camp must also be subdivided. Some of those chemists took as their ideal mathematical physics of the sort exemplified by Isaac Newton's *Principia*. Such reductionists, or *physicalists*, were dominated by the search for ultimate elements but tended to focus their attention on microscopic forces rather than on the presumed ultimate particles themselves. Opposed to this tradition were the nonreductionists, or *materialists*, who viewed the physicalists' search for the ultimate particles and forces as premature, hence sterile, and sought more proximate, cautious, and qualitative explanations for chemical phenomena. Those scientists focused on the sensible properties of the various kinds of matter.[6]

A useful way to distinguish these species of theorists is to identify their various scientific ideals. Both categories of conventionalists tended to regard chemistry as a branch of natural history. The Linnaean search for natural taxonomies of plants and animals could be applied directly to crystal forms, boiling-point and melting-point series, molecular formulas, and all other properties of homogeneous substances. Physicalists, on the other hand, attempted to consummate the Newtonian dream of extending an abstract mathematical program from celestial mechanics down to the submicroscopic realm of chemistry. The materialists' ideals were more akin to scholastic forms than to seventeenth-century mechanical philosophy; they sought concrete rather than abstract representations of the microcosm.

The above classification assists in rationally sorting out the myriad scientific disputes of the nineteenth century. But I would not want to force a very complex history into a Procrustean taxonomic bed. Indeed, historical examples demonstrate that few chemists can be classified in a completely unambiguous fashion. Instead of rigid orthodox conventionalists fighting tooth and nail with commonsensical if somewhat naive materialists, we shall find individual personalities in which these philosophical positions are blended and alloyed, indeed, frequently vying for dominance in the same scientist.

In general terms, physical science in the seventeenth century and the first half of the eighteenth century was dominated by a physicalist style of thought, whereas in the second half of the eighteenth century, materialism became more fashionable.[7] By contrast, the nineteenth century was far more interesting, because no single style was dominant. Rather, the various *Weltan-*

schauungen combined and competed in fascinating permutations and contextures. Indeed, they still seem to do so today.

This complexity of views can be seen at the beginning of the nineteenth century with the emergence of the chemical atomic theory—the theoretical basis of all modern chemistry, organic and inorganic alike. Despite the hegemony of the materialist style in the 1780s and 1790s, physicalist chemists had never given up the cherished Newtonian dream of deducing the ultimate elementary particles and forces of nature. But over the half-century following Newton's death, that approach had proven discouragingly sterile.[8] Antoine Laurent Lavoisier applied materialist precepts in defining a chemical element proximately and operationally as "the last point reached by chemical analysis" and in defining chemical analysis quantitatively and gravimetrically. By these definitions he arrived at the first recognizably modern table of the elements. But he was still enough of a physicalist to deny those elements ontological status as ultimate building blocks of nature. Nature *must*, he thought, be simple and unified. Some three dozen apparent elements must actually be composed of a much smaller number of true elements—say, one to three. Hence Lavoisier considered carbon, oxygen, nitrogen, sulfur, iron, cobalt, and so on, merely provisional elements reflecting the present inadequacies of chemical analysis. The number of such elements was expected to decline steadily as analytical methods continually improved.[9]

The son of a Quaker weaver, the English scientist John Dalton was much more fully committed to a materialist view. He transformed the provisional elements of the great French aristocrat into ontological elements, whose atoms he conceived to be essentially irreducible. This concept was really the *psychological* key to the chemical atomic theory.[10] The other prerequisite for the emergence of chemical atomism was a set of phenomena for which the new theory could provide satisfying explanations and thus prove its worth. This prerequisite was fulfilled by the discovery of the laws of stoichiometry between 1790 and 1810 by the Frenchman Joseph Louis Proust, the German Jeremias Benjamin Richter, and the Englishman Dalton. These laws were then verified with scrupulous precision by the great Swedish chemist Jakob Berzelius, who earns at least as much credit for the defense, diffusion, and elaboration of the atomic theory as Dalton earns for its creation.

Like Lavoisier, Berzelius was a complex mixture of physicalist and materialist tendencies. He followed Lavoisier's empiricism, accepted most aspects of Dalton's atomic theory, and propounded a distinctly materialist view of heat, electricity, and electrochemistry. On the other hand, like his great rival Sir Humphry Davy, he often assumed complexity in the elementary atoms in order to seek a greater simplicity in the subatomic realm. For example, chemical analogies led both chemists to look for hydrogen in the ammonium radical and in the metals, and oxygen in chlorine and nitrogen. Ironically, it was Davy—certainly more of a physicalist than Berzelius—who first gave up the search, declaring chlorine an element in 1810; ten more years passed before Berzelius was willing to concede the point.[11]

Berzelius' physicalist side was also revealed in his search for "rational formulas." He used this term to indicate those formulas that go beyond the empirical and into the theoretical realm, revealing the chemist's conception of the relative groupings of the atoms that constitute the molecule. Although this term applied to inorganic, as well as organic, chemistry, in the latter field the concept had by far its greatest application, particularly after the discovery of isomerism during the 1820s.

The concept of rational formulas was fundamentally physicalist because the physical arrangement of the atoms was considered a more important determinant of properties than the material nature of the atoms themselves. Isomers were immediately understood as compounds with identical empirical formulas but differing rational—we would say structural—formulas. But could the scientist ever infer anything that was both specific and reasonably well verified about those differing structures? Organic chemists of a century and a half ago accepted this transdictive challenge and charged bravely into battle. Although the victory was not easily won, the remarkable overall success of organic chemists during the middle decades of the nineteenth century transformed their specialty from a relatively obscure branch into the dominant field of chemistry, established the scientific basis of what became an enormous industry, and served as a model of transdictive success that provided inspiration to inorganic chemists, physical chemists, and physicists.

But let us return to problems faced by organic chemists *circa* 1830. A particular nexus of controversy was one of the basic organic molecules, ethyl alcohol.[12] It was known that the alcohol could be dehydrated by concentrated sulfuric acid to form ether or, under more stringent conditions, ethylene. Dumas believed that alcohol was therefore ethylene combined with two water molecules; loss of one by dehydration led to ether, loss of both generated ethylene. For other reasons, Berzelius formulated alcohol as a C_2H_6 radical combined with one oxygen atom. The German chemist Justus Liebig, who had established a close personal relationship with Berzelius after their first meeting in 1830, formulated alcohol as hydrated ether, ether itself being ethyl oxide, $C_4H_{10}O$. Similar disagreements were typical during the 1830s and 1840s.

These scientists, however, shared the same goal: breaking up compounds schematically into their component radicals. This endeavor had just gotten under way when a number of chemists embarked on a route that appeared better to them but seemed a time-wasting detour to others. Between 1834 and 1854 Dumas and a former student, Auguste Laurent, explored the phenomenon of chlorine substitution in organic molecules, leading to the formulation of Dumas' "type" theory and Laurent's "nucleus" theory. Like the Liebig-Berzelius radical concept, both new theories were transdictive, but cautiously so. This phenomenon of substitution seemed to strengthen the physicalists' hand as much as isomerism had, since the highly electronegative chlorine could replace the highly electropositive hydrogen, often with very minor alterations of chemical properties. In other words, once more it was arrangement, rather than the kind of atoms involved, that determined properties.

But one could go too far with this conviction. Friedrich Wöhler demonstrated this possibility with his renowned tongue-in-cheek report of cotton cloth composed of 100 percent chlorine gas, produced by successive replacement of every atom in each cotton molecule by chlorine, without altering the molecular structure. Concluding his report with the comment that this new synthetic fabric was all the rage among the London couturiers, he signed it S. C. H. Windler ("swindler" in German). Wöhler had written the satire as a private joke in a letter to his close friend Liebig, but Liebig published it for all the world to see in his *Annalen der Chemie.*[13]

By lampooning the ontological transdictive theories of Dumas and Laurent, Liebig was in effect retreating to a position closer to orthodox conventionalism. Indeed, precisely at the time of publication of the S. C. H. Windler critique, in early 1840, Liebig suddenly abandoned theoretical chemistry for the study of agriculture and physiology. For some time he had been becoming increasingly frustrated, or to use his precise expression, disgusted, by the intrinsic uncertainties and the mercurial character of organic chemical theory. In 1834 he complained to Berzelius that "the loveliest theories are overthrown by these damned experiments; it's no fun at all being a chemist anymore." Five years later Liebig declared that he had developed a "real fear of theoretical discussions."

> I envy woodcutters and office workers; just imagine the pure and untroubled joy these people feel when their day's work is done and they go home to enjoy the rest they deserve. Their minds are tranquil, their appetites ravenous, their sleep sound and carefree. How happily and with what comfort they must eat their sandwiches and drink their wine on Sunday picnics in the park; what do they care if croconic acid belongs in the canine or in the amphibian species; they have no other care than if the butter is rancid or the wine has gone bad. God, how I envy these happy souls![14]

Finally, in 1840, Liebig reported to Berzelius: "I have become totally sober—colder and more rational than you can imagine. . . . I was cured, it was an emetic, everything is disgorged and purged, and I am resolved never to mention it in my journal."[15] One of the most imaginative and prolific theorists of the

1830s had simply lost the faith. Liebig spoke no longer of atoms but now solely of "equivalents"; he even rejected his own hydracid theory, proposed as recently as 1838, and returned to the older dualistic oxyacid formulations. There is a sharp irony here, since the equivalent weights he embraced in the 1840s turned out to be not a whit more empirical than any rival set of atomic weights; and the oxyacid formulations were based on nothing other than an earlier theoretical conception that, moreover, we would today regard as being considerably less successful.

Liebig wrote to Berzelius concerning his conviction, shared by other chemists, that the concepts of atoms and atomic weights should be abandoned and replaced by the purportedly empirical equivalent weight concept. Liebig thought a formal proposal of this sort should come from the master himself, the dean of European chemists, Berzelius. But Berzelius, ever the ontological theorist, flatly refused. He correctly perceived the fallacy in regarding conventional equivalents as more empirical than atomic weights: "The concept of equivalents is relative to a specific series of compounds, and would be quite sufficient if bodies only combined in a single proportion. But since this is not the case, it is no longer positive, but merely conventional."[16]

The new preference for positivistic formulations is also detectable in the work of Charles Gerhardt, Laurent's friend and collaborator. Gerhardt went so far as to deny that chemists could ever learn anything at all about the internal arrangement of atoms that form molecules, thus representing a position close to orthodox conventionalism. For this view he received more than one demur from Laurent, who was much more of a methodological conventionalist. Gerhardt's goal was a rational taxonomy of organic compounds. Laurent objected. "Your classification is bad. . . . Without a dominating idea, it is impossible to do anything. Will you ever get anything from your classification? No, nothing, because there is no idea there. A classification must show a series of relationships. And I am persuaded that, whatever may be the point of departure, one will always be able to come to interesting relationships. But this point of departure must be an idea. . . . It is well and good to keep repeating that we need neutral ground on which the whole world can meet. Well! Good grief! How about alphabetical order?!"[17] For their iconoclasm and independence, for their defense of unpopular concepts and their seeming arrogance, Gerhardt and Laurent became the bêtes noires of European organic chemistry of the 1840s and early 1850s.

One might think that Gerhardt's orthodox conventionalism would have fit in well with the trend of the times, and so at first glance his status as pariah seems a bit curious. There are ironies aplenty here. First, from his conventionalist taxonomic starting point, Gerhardt proposed atomic weights nearly identical with those of the ontological theorist Berzelius. Second, neither Gerhardt nor Berzelius recognized the similarity of their weight theories, and each attacked the other's system ferociously. Third, when Gerhardt first made his proposal in 1842, the Berzelian hegemony still held sway, but after three or four years most European chemists had left both Gerhardt and Berzelius behind, embracing instead the Wollaston-Gmelin equivalents. And fourth, within a decade of Gerhardt's proposal an ontological theory arose on the basis of Gerhardt's conventionalist reform, and within two decades this ontological theory had recaptured the allegiance of most working chemists, especially organic chemists, and brought them back to the atomist fold.

Gerhardt is the chemist traditionally associated most closely with this "newer" type theory, as distinguished from Dumas' older theory of the same name. Between 1851 and his untimely death in 1856 Gerhardt did indeed develop the theory, in a fashion consistent with his taxonomic-conventionalist orientation and on the basis of his 1842 atomic weight reform. But another scientist truly deserves pride of place in this episode—the English organic chemist Alexander Williamson.

Williamson was the first prominent convert to the Gerhardt-Laurent reform, and in 1850 he was the *only* adherent of their ideas outside of France. Despite his personal association with Gerhardt during the 1840s and some private instruction in mathematics from none other than August Comte, Williamson somehow escaped the influence of the positivist school and acquired a distinct ontological style. In 1850 he discovered and published his celebrated ether synthesis, and in subsequent papers he did not hesitate to express his belief that in the ether molecule the oxygen atom physically links the two ethyl radicals together. Such statements clearly referred to the actual physical and chemical structure of the molecule and would have been anathema to a positivist. Furthermore, any entity that linked two units together could not possibly be a featureless, spherical, indivisible atom exerting isotropic forces. In modern vocabulary, atoms exhibiting valence absolutely require a more sophisticated model than the traditional billiard-ball image of the Daltonian atom.

Gerhardt created his much more general theory of types primarily on the basis of Williamson's ideas and experiments. But in doing so, Gerhardt transformed Williamson's ontological theory into a taxonomic-conventionalist theory. Still, it was the type theory in Williamson's less generalized but more concrete form that began to attract adherents during the 1850s and that ultimately led, toward the end of the decade, to the theories of atomic valence and molecular structure.

Parallel to Gerhardt's generalization of Williamson's type theory in a taxonomic-conventionalist direction was August Kekulé's generalization of the proposal in an ontological direction—a direction much more consistent with Williamson's original intent. Kekulé received his Ph.D. degree at Giessen under Liebig's direction but was also much influenced by his personal association with Dumas, Gerhardt, and Williamson in the early 1850s. Williamson had spoken of the oxygen atom, and other divalent atoms and radicals, "holding together" monovalent atoms and radicals in a single molecular unit. In the late 1850s Kekulé developed an analogous conception of, for example, trivalent nitrogen and tetravalent carbon. He further pointed out that if carbon atoms could hold other carbon atoms together—that is, if carbon atoms could form chains—then the detailed molecular structure of many well-known organic compounds could be specified immediately. This insight represents the creation of the theory of valence and its application to organic compounds in the theory of chemical structure. Both theories are prime examples of transdiction, and both require an ontological orientation for their creation and elaboration. Kekulé's widely-known anecdotes about his dreams of molecular chains and snakes, and his much-maligned sausage formulas, are evidence for his ontological-realist approach to science.

But at this point an important qualification requires some discussion. Williamson and Kekulé had a virtually identical theoretical style, but the precise classification of that style eludes me and my taxonomy. It tended toward ontological realism but was neither physicalism nor materialism; moreover, it retained elements of methodological conventionalism. One might suppose that both men were influenced by Gerhardt and his style and also by the prevailing positivistic trend of the day. Williamson, for instance, had a well-known aversion to complicated systems of notation, especially that of Hermann Kolbe, which seemed to imply structural information about compounds that did not emerge from the empirical data. He pointed out that, in writing a formula for acetic acid, Kolbe had used five different symbols or conventions, implying the operation of five different but unnamed chemical forces.

> It would be just as reasonable to describe an oak-tree as composed of the blocks and chips and shavings to which it may be reduced by the hatchet, as by Dr. Kolbe's formula to describe acetic acid as containing the products which may be obtained from it by destructive influences. A Kolbe botanist would say that half the chips are united with some of the block by the force *parenthesis;* the other half joined to this group in a different way, described by a *buckle;* shavings stuck on to these in a third manner, *comma;* and finally, a compound of shavings and blocks united together by a fourth force, *juxtaposition*, is joined on to the main body by a fifth force, *full stop*. The general use of unmeaning signs has become so habitual to Dr. Kolbe, that whenever anything has to be explained, he performs the task to his own satisfaction by inventing a sign for its unknown cause. . . . Signs and words are doubtless indispensable means for the expression of facts of thoughts; but Dr. Kolbe uses them instead of facts, and as a substitute for ideas.[18]

In a parallel way, the writings of Nicholas Fisher have in recent years illustrated the large extent to which Kekulé's structure theory emerged from purely taxonomic criteria, and the
writings of G. V. Bykov have stressed that many of Kekulés statements reflect the conventionalism of Gerhardt. "Rational formulas," Kekulé wrote in his textbook, "are formulas expressing
chemical reactions"—nothing more, nothing less.[19] Some historians have even denied that Kekulé's structure theory was intended to make any kind of statement at all concerning the internal arrangement of atoms in a molecule.

To argue in this fashion requires a much broader reading of
Kekulé's conventionalist statements than what I believe was intended, and it requires ignoring a great deal of contextual information that illustrates Kekulé's interest in, indeed his passion-
ate devotion to, ontological theory. For example, Kekulé always
argued that attempts at three-dimensional representation of organic molecules were still premature, since we had no solid information on the subject. Kekulé even devised a tetrahedral
model for the carbon atom but used it only for its convenience,
not for its heuristic power. When his famous student Jacobus H.
van't Hoff used a similar model to explain certain cases of isomerism, Kekulé continued to remain aloof. But he was utterly
convinced that the bonding order of carbon atoms in the skeleton and the positions of functional groups with respect to the
skeleton could be adequately established for most organic compounds. Many cases of what is now called structural or skeletal
isomerism could be explained using this relatively straight-
forward concept of chemical structure. Limited in this way,
Kekulé's theory was still sufficiently general and powerful to be
enormously successful and influential and, in the vocabulary of
this paper, both ontological and transdictive.

In short, both Kekulé and his friend and mentor Williamson
were extremely careful to draw a sharp distinction between
what we know and what we do not yet know, or more precisely,
between what had attained an adequate degree of verification
and what still remained more-or-less probable speculation. In

this regard the two men were equally critical of chemists who, like Gerhardt, were too cautious in their positivistic approach to explanation, and chemists who, like Kolbe, were presumptuously incautious in their specifications of "absolute constitutions" of organic molecules.

I have argued that Kekulé's ontological approach was a prerequisite for the creation of the theory of structure. But I would also argue that his tendency toward methodological conventionalism was highly useful. Supporting this thesis is the failure of all familiar models to explain the phenomenon of chain formation. Throughout the first half of the nineteenth century, chemists had tended to invoke one of two force models to explain the cohesion of chemical molecules: gravitation or electrical attraction. Gravitational models had been used, though never developed in detail, by Claude Louis de Berthollet, Dumas, and a few others; electrical attraction was considerably more popular and was stressed by chemists such as Davy and Berzelius. But neither force could readily be incorporated into a model that was capable of explaining the phenomenon of valence, much less the property of chain formation between identical atoms. The creation of the structure theory thus required the relinquishment of every familiar model in order to explain the nature of chemical affinity. Gravitational and coulombic forces would, of course, eventually reenter atomic chemistry and physics, but during the last half of the nineteenth century, the most fruitful and appropriate attitude toward this crucial question was the positivists' creed, Hypotheses non fingo.

Because the highly successful theories of valence and structure were a product of the Williamson-Kekulé school, whose reasoning was founded on the reformed atomic weights and formulas of Gerhardt and Laurent, many contemporary chemists regarded that reform to have thereby acquired crucial evidentiary support previously missing. Chemists thus began to return to atomic weights from their period of exile in the land of equivalents. When the international Karlsruhe Congress of 1860 passed a resolution to use henceforth only the weights of Gerhardt and Berzelius, it merely ratified a decision that had already essentially been made by the European chemical community. In sum, the maturation of the field of organic chemistry represented by the emergence of structure theory had wide significance for chemistry as a whole, in providing an effective rationale for the first general consensus regarding a single system of atomic weights and chemical formulas. Further, structure theory influenced all subsequent nineteenth-century physical science by providing the model of a highly successful transdictive ontological theory.

But this success by no means signaled a final victory of ontological over conventional theory. For another half-century certain scientists continued to cherish the dream of a positive science independent of microscopic models and mechanisms, one based solely on nontransdictive inference involving directly sensed, or macroscopic, phenomena. In physics this dualism is well exemplified by the theories of heat and gases. During the middle decades of the nineteenth century, the caloric theory, a transdictive materialist ontological theory, gave way to the emergent science of thermodynamics. Even today thermodynamics can be expressed with no reference whatever to microphysical models, and thus it conforms well with conventionalist precepts. An essentially modern formulation of thermodynamics was first enunciated in 1850 by Rudolf Clausius. Ironically, during the next decade Clausius himself, along with James Clerk Maxwell and others, developed an alternative approach called the kinetic theory, an approach not inconsistent with thermodynamics but distinctly transdictive and ontological. Like structure theory, the kinetic theory celebrated a number of striking successes in the following quarter-century despite continuing opposition by a significant school of physicists who hoped to see the more positive thermodynamic approach eventually devour the unappetizing kinetic theory.[20]

Similarly, positivist chemists continued to stage organized resistance to valence and structure theories. The most highly touted alternative was Benjamin Brodie's operationalist "chemical calculus," which purported to free chemistry from the pathos and bathos of atoms and molecules. Brodie found the works of

Kekulé and his French follower Alfred Naquet "scribbled over with pictures of molecules and atoms, arranged in all imaginable ways, and if there was no reason for this, it was a mischievous thing to do, for it led to a confusion of ideas, and to mixing up fictions with facts." Brodie's broadside attack evoked some sympathy in the chemical community. After the first oral presentation of his ideas to the Chemical Society, Williamson's good friend and associate William Odling ridiculed the concept of valence. "We have been led to believe that not only have we atoms, but that these atoms possess imaginary prongs, and that there is an imaginary clasping between them by means of these imaginary prongs, in a sort of hermaphroditism which it is scarcely possible to refer to."[21]

Even ontological theorists often found it appropriate to heap scorn on those ontological theories with which they disagreed. When van't Hoff, then an unknown young Dutchman working in the chemistry department of a veterinary school, extended Kekulé's structural concepts into three dimensions and thus created the basis for modern stereochemistry, Hermann Kolbe's reaction was a derisory howl. "A certain J. H. van't Hoff of the veterinary school of Utrecht," he wrote, "finds, so it seems, no taste for exact chemical investigation. He has thought it more convenient to mount Pegasus (obviously loaned by the veterinary school) and to proclaim in his Spatial Chemistry how, during his bold flight to the top of the chemical Parnassus, the atoms appeared to have grouped themselves in cosmic space."[22]

French skeptics, led particularly by Marcellin Berthelot, were also active in the last third of the century. Berthelot was a radical and an atheist, and it is not surprising that he remained agnostic about the existence of atoms, thus representing a classic model of orthodox conventionalism. He once told Naquet, "I do not want chemistry to degenerate into a religion; I do not want the chemist to believe in the existence of atoms as the Christian believes in the existence of Christ in the communion wafer."[23]

This movement of skeptical chemists came to a climax with the rise of "energetics," in Germany during the 1890s. Led by the physical chemist Wilhelm Ostwald, energeticists took as their ideal a thoroughly conventionalized thermodynamics and opposed all forms of atomism. Ostwald labeled mechanist-materialist approaches to science "hypothetical, even metaphysical"; indeed, "it is pure and simply an error. . . . I have been told that energy is after all an invention, an abstraction, whereas matter is real. I reply: quite the opposite! It is matter which is imaginary, an entity that we have rather imperfectly constructed to represent that which persists through the flux of appearances."[24]

But it would be a mistake to state that positivism was the leading scientific philosophy of the late nineteenth century. In fact, ontological theory in the form of various sorts of atomism was alive and well during this period, especially chemical atomism as it was used by organic chemists. For instance, Williamson himself defended the atomic theory with vehemence and great effectiveness in the 1860s and 1870s. He, Maxwell, and others pointed out that Brodie's supposedly hypothesis-free chemical calculus actually concealed assumptions that were equivalent to those made by chemical atomists. Similarly, all the defenders of conventionalist chemical theories, including such leading personalities as Berthelot and Ostwald, continued to utilize chemical atomism, though of course they drew little attention to this fact. Ostwald and Odling lived long enough to see the triumph of atomism in the first decade of the twentieth century, and both then publicly conceded the error of their earlier stance. Odling averred that he had been wrong to question Williamson on this point; he said that he had always been content to follow in Williamson's footsteps but had unfortunately more than once lagged behind.[25]

Ostwald referred repeatedly to the loneliness of his crusade for energetics. He was opposed by most of the leading *fin de siècle* physicists, especially Max Planck and Ludwig Boltzmann. One of the most effective rebuttals used against Ostwald by Boltzmann was taken not from physics but from organic chemistry: namely, the heuristic power of structural explanations for isomerism, especially stereoisomerism. In 1904 the empirical-minded chemist Ida Freund commented that the study of isomerism "has supplied the most striking proof of the validity and utility of the introduction into the science of these hypothetical magnitudes, the atom and the molecule."[26]

I would like to make two concluding comments. First, as much as I may appear to be awarding white and black hats to certain theoretical schools and especially to certain of their representatives, I would like to affirm my conviction of the fruitfulness of a variety of theoretical approaches. Such ontological theorists as Kolbe, who appears foolish to the modern reader in attacking the unknown young van't Hoff, made truly fundamental contributions to the science of chemistry in the nineteenth century.[27] By the same token, even the most seemingly arid positivists made similar fundamental contributions. The philosophy of Ernst Mach, who was the guiding spirit of the energetics movement, provided the point of departure for Einstein in developing his theory of relativity, in precisely the same sense that the positivist Gerhardt paved the way for Williamson and Kekulé.

Second, I simply want to underline once more the importance of Williamson's 1850 ether synthesis. It was this achievement, and the working out of its theoretical implications over the next decade, that led to the first convincing empirical chemical evidence for a set of atomic magnitudes on which all could agree. In a more general sense, the ether synthesis also led to the emergence of valence and structure theory, especially at the hands of Williamson's protégé August Kekulé. For the remainder of the century, whatever doubts were expressed by conventionalists were not shared by most organic chemists, who recognized that the ontological roots of atomic and structural theory were solidly planted. They minded their own business, applied their theories with a sense of unquestioning confidence, and were daily rewarded for their faith.

NOTES

1. Maurice Mandelbaum, *Philosophy, Science and Sense Perception* (Baltimore, 1964), 245.
2. J.-B. Dumas, *Leçons de philosophie chimique* (Paris, 1837), 60.
3. Mandelbaum, *Philosophy, Science and Sense Perception;* Alexandre Koyré, "Galilée et la loi d'inertie," *Etudes galiléennes* (Paris, 1939).

4. Pierre Duhem, *The Aim and Structure of Physical Theory*, trans. P. P. Wiener (Princeton, 1954), esp. 7–30.
5. For discussions of specific examples of nineteenth-century conventionalist physical theory, see E. N. Hiebert, *The Conception of Thermodynamics in the Scientific Thought of Mach and Planck* (Freiburg, 1968); Gerald Holton, *Thematic Origins of Scientific Thought* (Cambridge, Mass., 1973); and Mary Jo Nye, "Berthelot's Anti-Atomism: A 'Matter of Taste'?" *Annals of Science*, 38 (1981), 585–90.
6. Physicalism and materialism are discussed by J. K. Bonner, "Amedeo Avogadro" (Ph.D. dissertation, Johns Hopkins University, 1974); Robert Fox, "The Rise and Fall of Laplacian Physics," *Historical Studies in the Physical Sciences*, 4 (1974), 89–136; Robert Schofield, *Mechanism and Materialism: British Natural Philosophy in an Age of Reason* (Princeton, 1970); and Arnold Thackray, *Atoms and Powers: An Essay on Newtonian Matter-Theory and the Development of Chemistry* (Cambridge, Mass., 1970).
7. This is a major thesis of Schofield's *Mechanism and Materialism*.
8. See Thackray, *Atoms and Powers*.
9. R. Siegfried and B. J. Dobbs, "Composition, a Neglected Aspect of the Chemical Revolution," *Annals of Science*, 24 (1968), 275–93; R. Siegfried, "Lavoisier's Table of Simple Substances," *Ambix*, 29 (1982), 29–48.
10. Arnold Thackray, *John Dalton: Critical Assessments of His Life and Science* (Cambridge, Mass., 1972), 61–88.
11. Alan J. Rocke, *Chemical Atomism in the Nineteenth Century* (Columbus, Ohio, 1984), 153–93.
12. *Ibid.*, 174–76, 216.
13. F. Wöhler [S. C. H. Windler], "Ueber das Substitutionsgesetz und die Theorie der Typen," *Annalen der Chemie und Pharmacie*, 33 (1840), 308.
14. J. Carriere (ed.), *Berzelius und Liebig: Ihre Briefe von 1831–1845* (2nd ed.; Munich, 1898), 94, 191.
15. *Ibid.*, 210–11.
16. *Ibid.*, 206.
17. Marc Tiffeneau (ed.), *Correspondance de Charles Gerhardt* (2 vols.; Paris, 1918–25), Vol. 1, pp. 5, 19–20. N. W. Fisher provides a perceptive analysis of the relations between the two French chemists in "Organic Classification Before Kekulé," *Ambix*, 20 (1973), 215–17.
18. Alexander Williamson, "On Dr. Kolbe's Additive Formulae," *Journal of the Chemical Society*, 7 (1854), 123, 132–35.
19. N. W. Fisher, "Organic Classification Before Kekulé," and "Kekulé and Organic Classification," *Ambix*, 21 (1974), 29–52; G. V. Bykov, "The Origin of the Theory of Chemical Structure," *Journal of Chemical Education*, 39 (1962), 220–24; R. Anschütz (ed.), *August Kekulé* (2 vols.; Berlin, 1929), Vol. 2, p. 112.
20. Stephen G. Brush, *The Kind of Motion We Call Heat* (2 vols.; Amsterdam, 1976); Stephen G. Brush, *The Temperature of History* (New York, 1977).
21. "Discussion on Dr. Williamson's Lecture on the Atomic Theory, November 4th, 1869," *Journal of the Chemical Society*, 22 (1869), 440, 436. See also William Brock (ed.), *The Atomic Debates* (Leicester, U.K., 1967), 19–26.
22. H. Kolbe, "Zeichen der Zeit," *Journal für praktische Chemie*, 15 (1877), 473.
23. C. Graebe, "Marcellin Berthelot," *Berichte der Deutschen Chemischen Gesellschaft*, 41 (1908), 4855.

24. W. Ostwald, *Abhandlungen und Vorträge allgemeinen Inhaltes* (Leipzig, 1904), 224, 226, 230, 234.
25. *Chemical News*, 78 (1898), 286.
26. Ida Freund, *The Study of Chemical Composition* (Cambridge, U.K., 1904), 559.
27. Kolbe's reputation has been resurrected by Henry Armstrong, "The Doctrine of Atomic Valency," *Nature*, 125 (1930), 807–10, and by Harold Hartley, *Studies in the History of Chemistry* (Oxford, 1971), 195–222.

JOHN H. WOTIZ and SUSANNA RUDOFSKY

The Unknown Kekulé

CHEMISTS and nonchemists, psychologists in particular, are likely to know Friedrich August Kekulé not only as one of the founders of structural organic chemistry but also as a scientist who conceived some of his ideas in dreams. A careful scrutiny of the traditional story that has evolved around Kekulé indicates that there is room for questions on the matter of his visions.

Kekulé was born in 1829 in Darmstadt, Germany. He started studying architecture at Giessen University in 1847 but, under the influence of Justus Liebig in Giessen, abandoned it for the study of chemistry in 1849. During 1851 and 1852 he studied in Paris, where he associated with Jean-Baptiste Dumas, August Laurent, and Charles F. Gerhardt. After receiving his Ph.D. degree from Giessen University in 1852, he accepted the position of private assistant, first, to Alfred von Planta in the Schloss Reichenau near Chur, Switzerland, from 1852 to 1853, and next, to J. Stenhouse in London from 1854 to 1855. During his London stay he associated with Alexander W. Williamson, Edward Frankland, and William Odling at the University College. His habilitation, in 1856, was at the University of Heidelberg, where Kekulé served as a privatdozent from 1855 to 1858. His first professorship started in 1858 at the newly established, state-supported University in Ghent, Belgium. From 1867 until his death in 1896 he was a professor at Bonn University.

We would like to thank Klaus Hafner, Director, and Mrs. Susanne Priebe, librarian of the Kekulé Archives, located at the Technical University in Darmstadt, West Germany, for their assistance and cooperation in providing pertinent letters and documents, some previously unpublished.

21

The following passage, from a widely used undergraduate text-book on organic chemistry, reflects the widespread misconceptions surrounding Kekulé.

In 1858, August Kekulé (of the University of Bonn) had proposed that carbon atoms can join to one another to form *chains*. Then, in 1865, he offered an answer to the question of benzene: these carbon chains can sometimes be closed, to form *rings*. "I was sitting writing at my textbook, but the work did not progress; my thoughts were else-where. I turned my chair to the fire and dozed. Again the atoms were gamboling before my eyes . . . all twisting and turning in snake-like motion. But look! What was that? One of the snakes had seized hold of its own tail. . . . As by a flash of lightning I woke. . . . I spent the rest of the night working out the consequences of the hypothesis. Let us learn to dream gentlemen, and then perhaps we shall learn the truth."—August Kekulé, 1865.[1]

The citation incorrectly places Kekulé in Bonn in 1858 as a professor, and the textbook gives the impression that the quotation is taken from Kekulé's first paper on the cyclic structure of benzene, published in 1865. However, the first reference to Kekulé's dreams appears in the proceedings of the Benzol Fest, a meeting in 1890 that celebrated the twenty-fifth anniversary of Kekulé's benzene-structure paper.[2]

In 1865 Kekulé considered the cyclic structure of benzene in two French publications. A somewhat expanded version appeared in a German publication in 1866, though it has been frequently listed incorrectly as published in 1865.[3] The origin of this error is an ambiguous statement in Kekulé's 1872 paper dealing with condensation products of aldehydes. In the paper, Kekulé reviewed the status of the "new aromatic" chemistry, stating that he had first published his views on the constitution of aromatic compounds in 1865 but referring by footnote to *Annalen der Chemie und Pharmacie*, Volume 137.[4] Kekulé's theory on the constitution of benzene appeared in print in 1865, but in the French publications; the German paper he cited appeared a year later. Kekulé made this statement one year after the end of the Franco-Prussian war and the beginning of a unified Germany under Prussian leadership. German nationalism was on the rise, and Kekulé was a professor at the relatively new Prussian university in Bonn. With the development of new attitudes, the Germans came to consider aromatic chemistry a unique national development; and the ambiguous handling of the citation implied the primacy of the German publication.

Several peculiar events occurred at the 1890 Benzol Fest and in the newspaper accounts of this celebration. Despite his initial reluctance, Kekulé agreed to be honored by the German Chemical Society and his friends and former students, who planned the commemoration. Because the anniversary of the publication date of the first benzene-structure paper coincided with the kaiser's birthday, it was agreed to hold the meeting on March 11, 1890, in the Berlin city hall. Chemists, educators, industrialists, and government representatives gathered from around the world to pay their respects and to read appropriate scientific papers. Kekulé, as intended, spoke last, without the aid of a manuscript. One week before the fest, Kekulé had written to Adolf Baeyer, one of the meeting organizers: "I do not plan to hold a long speech at all, I intend to leave this ultimately to the inspiration of the moment. . . . All the rest of you . . . can prepare talks. While I, poor man, have to rely almost entirely on improvisations."[5]

After the fest, August Wilhelm von Hofmann, president of the German Chemical Society, asked Gustav Schultz, an industrial chemist, to collect manuscripts from the speakers for publication in the society's journal, *Berichte der Deutschen Chemischen Gesellschaft*. Less than two weeks after the fest, Kekulé wrote to Schultz of his displeasure over this development. "Your letter shocks me more than the recent earthquake. What! Is the metropolis of Berlin such a crow's nest that not one stenographer was present at the celebration of 11 [March]? How will you be able now to achieve a faithful report? Now everyone will write to you, if he writes at all, what he wishes to have said; but not what he has said." (This may have become a self-fulfilling prophecy so far as Kekulé's report is concerned.) However, Kekulé promised to reconstruct his speech in spite of his "dulled memory," and he sent the manuscript to Schultz on April 7 with a self-deprecating and self-incriminating cover letter. "It is certainly difficult to say something rational at such an occasion . . .

the stupidities one has said. . . . I am always concerned of having terribly embarrassed myself . . . it would be better if one burned the whole rubbish and did not allow anything to be printed."[6]

An entire account of the proceedings appeared on May 5, 1890. In spite of his original intent, Kekulé wrote at length, devoting approximately 460 words to acknowledgment of honors that had been bestowed on him and about 3,340 words to personal reminiscences, advice to the younger generation, and poetry. In the reminiscences Kekulé introduced the passage about dreams with the following words: "Perhaps it will interest you if I let you know through highly indiscreet disclosures from my inner life, how I arrived at some ideas." He then wrote about the time he fell into a reverie on top of a bus in London and saw atoms gamboling before his eyes. The vision, which occurred sometime in 1854 or 1855, allegedly gave him the idea that carbon atoms can link together, forming chains. Kekulé claimed that a second dream, of a snake seizing its tail, occurred in Ghent, most likely in 1861 or 1862, while he was dozing in front of the fireplace. The result of the London carbon-chain dream was published in 1858, about four years after the dream supposedly occurred, and the Ghent paper allegedly inspired by the snake dream was published in 1865, about three years after it occurred. This pattern invalidates the popular notion that Kekulé, following a visionary dream, quickly wrote down his findings, and the Benzol Fest transactions contain his explanation (rationalization) of why he did not immediately publish his ideas.[7] Furthermore, there seems to be no evidence that Kekulé ever mentioned the dreams prior to the 1890 publication—a delay of twenty-eight and thirty-five years, respectively. Thus, if they actually occurred, the dreams were well-preserved secrets.

That Kekulé even mentioned the Ghent snake-benzene dream in his oral presentation is also difficult to believe, in spite of what he wrote after the fest. Our reservation is based on an examination of the newspaper accounts of the fest, which was well attended by the press. Of twenty-eight different articles, some printed in the Berlin papers the day after the fest, and others a few days later in the provincial and weekly publications, only three referred to dreams, and surprisingly, none of the articles mentioned the Ghent snake dream. For example, the *Berliner Tageblatt* devoted about 1,360 words to the fest but did not mention dreams. The *Tägliche Rundschau*, in an 800-word report, stated only that Kekulé "had seen the combining of atoms for the first time in London in a dream." According to the *National Zeitung*, "The honored guest told with whimsical humor of the dreams of his youth of the atoms that gamboled around him." The *Chemiker-Zeitung*, supposedly written by expert chemist-journalists for chemists and nonchemists alike, devoted about 1,600 words to the fest. It mentioned the London dream but apparently made two factual errors. "In London a dream image showed him the combining of atoms of benzene [?] for the first time and thus . . . he could arrive at this current concept of arrangements of atoms. When he awoke from the dream he sketched the basic features of the structural theory, examined and worked it out fully, and published it only 8 months [?] later." Another newspaper stated, "Thereafter Kekulé took the floor to express thanks in a fairly long speech for the greatly exaggerated honors and to demonstrate at the same time how he arrived at his theory in a completely natural manner." This characterization is interesting, since discoveries or revelations in dreams hardly occur in a "natural manner."[8] Furthermore, the absence of any mention of the snake-benzene dream in all the examined newspaper articles is noteworthy, because even in the nineteenth century reporters would not have passed over an opportunity to sensationalize such an account.

There is another good and simple reason for being skeptical of Kekulé's associating the self-devouring snakelike flames in a fireplace, resembling the ouroboros, with a cyclic structure for benzene. Laurent's book of 1854 contained a far more tangible precedent for a cyclic structure—a hexagon used to represent benzoyl chloride and ammonia (Fig. 1). We know that Kekulé was familiar with this book and the pages containing the hexagon, since he referred to the precise pages in his 1858 *Annalen der Chemie und Pharmacie* publication. He was favorably impressed by Laurent's book, as is evident from a letter he wrote to a German book publisher suggesting that he translate Laurent's book from French into German.[9] It is puzzling that Kekulé did not mention Laurent's book in his benzene-structure papers or in his Benzol Fest speech. Was he forgetful, or did he omit this reference on purpose?

Pour faire comprendre ce remplacement réciproque de deux restes, je supposerai que, dans l'ammoniaque et le chlorure de benzoïle, les atomes sont disposés suivant les figures hexagonales :

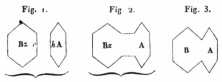

Fig. 1.   Fig. 2.   Fig. 3.

Bz et A, *fig.* 1, représentent le chlorure de benzoïle et l'ammoniaque, au moment où ils vont réagir l'un sur l'autre, l'arête *c* en face de l'arête *h* qu'elle doit enlever. Bz et A, *fig.* 2, représentent les deux restes pendant la réaction, et BA, *fig.* 3, les deux restes après la réaction, ayant comblé, réciproquement, les vides qui s'étaient formés dans A et dans B.

La *fig.* 3 représente une diaméride ou une figure qu'on peut diviser en deux parties A et B, mais à la condition de restituer à B l'arête soit de chlore, soit d'hydrogène qui lui manque, et de restituer également à A l'arête qu'il a perdue, ou bien une arête équivalente.

Fig. 1.   Hexagonal figures in Laurent's *Méthode de Chimie*, 1854.
Chemistry in Britain, 20 (1984), 721. Reproduced by permission.

The fest accounts of snake dreams also may be questioned. A clue lies in the account in the *National Zeitung:* "The honored guest told with whimsical humor of the dreams of his youth." Is it possible that Kekulé's whimsical and perhaps subtle humor was misunderstood or that it did not come through when he tried to reconstruct his extemporaneous remarks for the fest proceedings publication? Kekulé was indeed an individual who enjoyed the company of friends and associates. His first wife Stephanie (née Drory) died shortly after giving birth to their son

Stephan in Ghent in 1863. He married his housekeeper while in Bonn in 1876. This was an intellectual mismatch, and his home life was not all it might have been. (A noted Kekulé biographer described Luise, born Högel, as a "dragon." [10]) This circumstance drove Kekulé to seek relaxation with fellow chemists outside the home. Social drinking was practiced, but chemistry talk was the main objective. In 1886, at one such gathering, Kekulé was cleverly spoofed by his friends in a *Bierzeitung,* the *Berichte der Durstigen Chemischen Gesellschaft,* or the *Report of the Thirsty Chemical Society* (Fig. 2). The articles and the format of this fictitious journal were so like the publications in the *Berichte der Deutchen Chemischen Gesellschaft,* or the *Report of the German Chemical Society,* that even some chemists mistook it for a legitimate publication. The manuscript of the article by F. W. Findig (Inventive), entitled "About the Constitution of Benzene," was received on June 31 (June has only 30 days) and reported at an imaginary meeting on September 20, 1886, with Mr. Aujust Kuleke (*sic*) presiding. In this paper Findig described how one can formulate the benzene structure by visualizing a cage with six monkeys joined in a ring, holding each other by their hands and tails (Fig. 3). Findig pointed out that by switching handholds and tailholds the monkeys demonstrated the tautomeric shifts of single and double bonds of benzene structures. This clever spoof was especially effective in the original German, since it was based on a play on words: *Valence* is *Affinität* and *monkey* is *Affe.* It is reasonable to suppose that Kekulé tried to "get even" with the anonymous spoofers by suggesting that a self-devouring snake, not a cage of six monkeys, gave him the clue to the cyclic benzene. Other spoofs at the *Durstige Gesellschaft* session contained poetry readings, as did Kekulé's speech at the Benzol Fest. Obviously, the transcript of the fest did not record whether Kekulé's remarks were delivered with a grin so that the audience would realize that it was listening to an imaginary tale. However, the attending newspaper reporters knew well enough and omitted accordingly. [11]

The whimsical side of Kekulé comes through in other passages of his fest speech. Describing his London dream, he wrote: "I fell

BERICHTE

DER

DURSTIGEN

CHEMISCHEN GESELLSCHAFT.

UNERHÖRTER JAHRGANG.

No. 20.

(Ausgegeben am 20. September.)

BERLIN.

EIGENTHUM DER DURSTIGEN CHEMISCHEN GESELLSCHAFT
COMMISSIONSVERLAG von R. FRIEDLÄNDER & SOHN
N.W. CARLSTRASSE 11
1886.

Fig. 2. Title page of *Berichte der Durstigen Chemischen Gesellschaft*. Chemistry in Britain, 20 (1984), 722. Reproduced by permission.

Fig. 3. Monkey illustrations in *Berichte der Durstigen Chemischen Gesellschaft*. Chemistry in Britain, 20 (1984), 723. Reproduced by permission.

into a reverie, and lo, the atoms were gamboling before my eyes!
. . . I saw how frequently two smaller atoms . . . kept hold of
three or even four of the smaller; whilst the whole kept whirling
in a giddy dance. . . . I saw what our past master Kopp, my
highly honored teacher and friend, has depicted with such charm
in his 'Molekular-Welt'; but I saw it long before him." Hermann
Kopp's *Aus der Molekular-Welt*, published in 1881, is a humorous
description of the behavior of molecules. For example, Kopp described an imaginary club for molecules: "Purpose of the club:
dancing. . . . There they are dancing, the molecules! Now faster,
now less fast, depending on whether they are heavier or lighter,
they pass before our curious eye."[12] Kopp was one of Kekulé's
teachers at Giessen, and it is difficult to imagine that as a student Kekulé would have given his master the idea to visualize
atoms and molecules as dancing. Kekulé and Kopp convey essentially the same idea, the movement of particles in a humorous manner. Why Kekulé claimed priority for this "educational
tool" is again difficult to understand or even to rationalize.

Many of the present readers may have also been skeptical of
the accounts of Kekulé's dreams, but strangely enough, no one to
our knowledge has tried to document such "heresy." Distorted
and embellished interpretations of Kekulé's dreams are numerous, especially in the writings of psychologists. These citations
prompted Stephan Kekule von Stradonitz, Kekulé's son, to write
in 1927 in defense of the Kekulé myth that had been growing
ever since the 1890 fest.[13] Although no one had heard about the
dreams until they appeared in the fest proceedings, Stephan
claimed that the snake vision was mentioned by his father in
the "smallest circle of his family." No other reports verify this
statement.

Kekulé's claim to have originally conceived the benzene structure in a dreamlike revelation also seems inconsistent with a
passage that refers to his 1858 publication. "It now appears to
me opportune to publish the fundamental principles of a theory
which I conceived a rather long time ago, concerning the aromatic substance."[14] Which Kekulé are we to believe, the one who
claimed to have developed the aromatic theory in 1855 and pub-

lished it in 1858 or the Kekulé who claimed to have conceived
the benzene structure in 1861 or 1862 in a dreamlike revelation,
published it in 1865, and told about it twenty-five years later,
in 1890?

The notion that Kekulé's contributions were an act of genius
was created by his contemporaries in an effort to demonstrate
that aromatic chemistry had unique German foundations on
which the monopolistic German dye and pharmaceutical industry of the late nineteenth century securely rested.[15] The architect
of structural organic chemistry was no genius, but he was a
gifted opportunist. Some words from his fest speech clearly bear
this out. "I must first of all tell you that to me the benzene theory
was only a consequence, and a very obvious consequence of
the views that I formed about the valences of the atoms and of
the nature of their binding. . . . My colleagues! We all stand
on the shoulders of our predecessors; is it then surprising that
we can see further than they? . . . Everyone of my colleagues has
contributed to these developments, each in his own way . . . it
would be going too far to give me special credit for that [ben-
zene theory] development. Certain ideas at certain times are in
the air; if one man does not enunciate them, another will do so
soon afterwards." Archibald Scott Couper published independently in 1858 the two structural principles for which Kekulé
now gets credit.[16] Why did Kekulé fail to mention Couper in the
fest proceedings? After all, he praised by name Laurent, Dumas,
Jöns Jakob Berzelius, Frankland, Liebig, Heinrich Will, Charles
Adolphe Wurtz, Gerhardt, and Williamson. Furthermore, Couper formulated a cyclic structure for cyanuric acid in 1858, and
Laurent used hexagons for benzoyl chloride and for ammonia in
1854.[17] The mental picture resulting from Laurent's hexagon is
closer to the benzene structure than is the ouroboros (Fig. 4).

It is interesting to speculate about what prompted Kekulé to
write about the ouroboros. Attributing the conception of the
benzene structure to a dream, Kekulé reserved to himself a
unique priority of ideas. Rising German nationalism may also
have been a contributing factor. After all, using the tale of conceiving ideas in a dreamlike revelation, Kekulé did not have

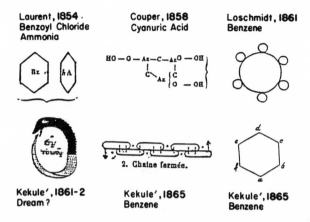

Fig. 4. Some early representations of cyclic structures.

to credit the foreign investigators—Laurent, Couper, and the Austrian Josef Loschmidt—on whose shoulders he was standing. In a letter to Hans Hübner on July 15, 1870, during the Franco-Prussian war, Kekulé referred to France as a nation of sons of bitches (Hundevolk diese Franzosen!).[18] During the war Louis Pasteur had returned the honorary Doctor of Medicine degree that Bonn University awarded him in 1868, a manifestation of Pasteur's French patriotism and an act of defiance against the brutalities of Prussian soldiers. Kekulé was a professor at Bonn University from 1867 on, and a possible linkage of his attitude to the Pasteur action requires further investigation.

It seems that Kekulé was a basically modest man who knew that he was only a link in the chain of progress and the evolution of chemistry. He comes through with all the failings of a man who exercised selective memory to improve his place in history. He certainly had the chance to affirm or to deny the notion of his dream-inspired genius by appropriate explanations. With the exception of a short note,[19] Kekulé's last publication appeared in 1890, and he died in 1896 after a short illness, at the age of 67. After a rich record of publications, why did he choose to remain silent for the last six years of his life? After all, many good investigations were completed in his laboratory—for example, on the structure of pyridine—but went unpublished. Was he embarrassed by the myth that started at the Benzol Fest, or was he unable or reluctant to confirm or deny it?

In 1895 Kekulé became Kekule von Stradonitz as a result of the research of his son Stephan, who documented his family's descent from an old Bohemian (Czech) noble family that had left Bohemia during the Thirty Years War because of their Protestant faith. The kaiser restored the title but, to the disappointment of the family, especially Stephan, did not attribute to Kekulé's contribution to chemistry the basis for that action.

Chemists and the general public have learned much about Kekulé from the two-volume (1,000 pages each) biography written by his student and colleague, Richard Anschütz. The biography was published to commemorate the 100th birthday of Kekulé. Anschütz, like Kekulé, was born in Darmstadt. He succeeded Kekulé in 1898 as professor at Bonn, and upon retirement moved many of Kekulé's belongings and much of his correspondence back to Darmstadt, with the Kekulé family's permission. This collection became the nucleus of the present Kekulé Archives at Darmstadt Technical University. In the biography, Anschütz reprinted many of Kekulé's pertinent publications, letters, and documents. Some researchers think that "since 1929 these volumes have been the most indispensable and reliable single guide for all Kekulé studies." We concur that the biography is a useful guide to Kekulé's work, but it is not always an accurate or reliable source of information.[20] For example, Anschütz reprinted the 1872 paper in which Kekulé referred ambiguously to his publication in *Annalen der Chemie und Pharmacie*, Volume 137, leaving the reader to surmise that it appeared in 1865 when in fact the paper was published in 1866. Although Anschütz added the correct date to Kekulé's footnote, he failed to acknowledge that it was he, not Kekulé, who had included it. In another instance, Anschütz, in Volume 1, page 658, arbitrarily omitted printing item 9 in Kekulé's *Maturitätszeugnis* and spelled Kekulé's name with an *accent aigu*, whereas the document did not contain the accent mark. Since on page 647, Volume 1, Anschütz explained that the accent mark was adopted by Kekulé's father, Ludwig Carl Emil, to avoid the mispronunciation by the French as "Kekyl," we do not understand why the biography has a picture of Kekulé's grandfather, Johann Wilhelm, identified as Kekulé. This posthumous change instituted by Anschütz should have been appropriately footnoted. Some letters—for example, the letters of Kekulé to Schultz—were not included in the biography, perhaps because they were embarrassing.[21] We wonder what the public opinion of Kekulé would be today if Anschütz had published the two self-incriminating letters. There is no substitute for searching the original literature and sources of information. We realize that we are dealing with another example that should have been an entry in the book *Betrayers of the Truth*.[22]

NOTES

1. Robert T. Morrison and Robert N. Boyd, *Organic Chemistry* (4th ed.; Boston, 1983), 574 (The quotation can also be found in the second and third editions). See also Frederic J. Kakis, "On Writing About Research," *Chemical and Engineering News*, September 29, 1975, p. 4. The author either used a very loose translation of, or paraphrased, Kekulé's statement.
2. Gustav Schultz, "Bericht über die Feier der Deutschen Chemischen Gesellschaft zur Ehren August Kekulé's," *Berichte der Deutschen Chemischen Gesellschaft*, 23 (1890), 1265–1312; Francis R. Japp, "Kekulé Memorial Lecture," *Journal of the Chemical Society* (London), 73 (1898), 97–138; O. Theodor Benfey, "August Kekulé and the Birth of the Structural Theory of Organic Chemistry in 1858," *Journal of Chemical Education*, 35 (1958), 21–23. Benfey gives the most recent complete translation of Kekulé's speech as reported by Schultz.
3. August Kekulé, "Sur la constitution des substances aromatiques," *Bulletin de la Société Chimique de Paris*, 3 (1865), 98–111, "Note sur quelques produits de substitution de la benzine," *Bulletin de l'Académie royale de Belgique*, 19 (1865), 551–63, and "Untersuchungen über aromatische Verbindungen," *Annalen der Chemie und Pharmacie*, 137 (1866), 129–96. See John H. Wotiz and Susanna Rudofsky, "Was There a Conspiracy When Kekulé's First German Benzene-Structure Paper Was Frequently Listed as Published in 1865?" *Journal of Chemical Education*, 59 (1982), 23–24.
4. August Kekulé, "Ueber einige Condensationsproducte des Aldehyds," *Annalen der Chemie und Pharmacie*, 162 (1872), 77–124.
5. John H. Wotiz and Susanna Rudofsky, "Kekulé's Dreams: Fact or Fiction?" *Chemistry in Britain*, 20 (1984), 720–23; August Kekulé to Adolf Von Baeyer, March 5, 1890, in Kekulé Archives, Darmstadt Technical University.
6. August Kekulé to Gustav Schultz, March 22, 1890, and April 7, 1890, both in Kekulé Archives. See Wotiz and Rudofsky, "Kekulé's Dreams," for a translation of the cover letter.
7. Schultz, "Bericht zur Ehren August Kekulé," 1265–1312; Benfey, "August

Kekulé and the Birth of the Structural Theory," 22; August Kekulé, "Ueber die Constitution und die Metamorphosen der chemischen Verbindungen und über die chemische Natur des Kohlenstoffs," *Annalen der Chemie und Pharmacie*, 106 (1858), 129–59; Wotiz and Rudofsky, "Kekulé's Dreams."

8. Wotiz and Rudofsky, "Kekulé's Dreams"; *Berliner Tageblatt, Tägliche Rundschau* (Berlin), *National Zeitung* (Berlin), all March 12, 1890; *Chemiker-Zeitung* (Cöthen), March 15, 1890; *Neue Preussische Kreutz-Zeitung* (Berlin), March 12, 1890; Byron J. Vanderbilt, "Kekulé's Whirling Snake: Fact or Fiction," *Journal of Chemical Education*, 52 (1975), 709.

9. Auguste Laurent, *Méthode de Chimie* (Paris, 1854), 408, 411; August Kekulé to Vieweg, a publisher in Braunschweig, July 4, 1854, in Kekulé Archives.

10. From a conversation of J. Gillis with J. Wotiz. Gillis (1893–1978) taught at Ghent University and wrote "Auguste Kekulé et son oeuvre, réalisée à Gand de 1858 à 1867," *Memoires, Académie royale de Belgique*, 37 (1966), 1.

11. Wotiz and Rudofsky, "Kekulé's Dreams"; John Read, *Humor and Humanism in Chemistry* (London, 1947), 221 (The author attributed the spoof to E. Jacobsen and O. N. Witt); August Kekulé to Vieweg, July 4, 1854.

12. Benfey, "August Kekulé and the Birth of the Structural Theory," 22; Hermann Kopp, *Aus der Molekular-Welt* (3rd ed.; Heidelberg, 1886), 42–44.

13. Susanna Rudofsky and John H. Wotiz, "The Impact of Kekulé's Dream Accounts on Psychologists and Psychoanalysts," *Abstracts of Papers, 186th American Chemical Society Meeting* (Washington, D.C., 1983), HIST 37; Stephan Kekule von Stradonitz, "Zwei chemische Visionen," *Angewandte Chemie*, 40 (1927), 736–37.

14. Kekulé, "Sur la constitution des substances aromatiques," 98.

15. Descriptions of Kekulé as a genius are found in, for example, the *Berliner Tageblatt*, March 12, 1890, and Schultz, "Bericht zur Ehren August Kekulé's," 1265–1312.

16. Benfey, "August Kekulé and the Birth of the Structural Theory," 21; A. Couper, "Sur une nouvelle théorie chimique," *Comptes Rendus Hebdomadaires des Séances de l'Académie des Sciences* (Paris), 46 (1858), 1157–60. James Kendall, *Great Discoveries by Young Chemists* (New York, 1953), 81–89 provides a good analysis of the circumstances which prevented Couper from claiming priority for conceiving the aromatic theory.

17. Couper, "Sur une nouvelle théorie chimique"; Laurent, *Méthode de Chimie*, 408.

18. August Kekulé to Hans Hübner, July 15, 1870, in Kekulé Archives.

19. August Kekulé, "Zur Kenntnis des Formaldehyds," *Berichte der Deutschen Chemischen Gesellschaft*, 25 (1892), 2435.

20. Richard Anschütz, *August Kekulé* (2 vols.; Berlin, 1929); Erwin N. Hiebert, "The Experimental Basis of Kekulé's Valence Theory," *Journal of Chemical Education*, 36 (1959), 320–27; Wotiz and Rudofsky, "Kekulé's Dreams."

21. Kekulé to Schultz, March 22, 1890, and April 7, 1890.

22. William Broad and Nicholas Waden, *Betrayers of the Truth* (New York, 1982).

From Molecular Morphology to Universal Dissymmetry

EARLY modern chemistry inherited from its precursor protoscience a number of general guiding concepts that, in specialized forms, gave shape to chemical classification procedures during the first half of the nineteenth century. The most important of these principles for theories of molecular morphology was the conception that reciprocal analogies in structure and function obtain between the large and the small. The theory that each microcosm epitomizes the macrocosm has an extended history, and it was not necessarily opposed to another traditional conception, that all substances are made up of two complementary and antagonistic active principles, the yin and the yang of Chinese alchemy, or the mercury and the sulphur of Arabian and European alchemy, based upon Greek antecedents.[1] During the early nineteenth century the two specialized derived traditions, the polar dualistic theory of Jakob Berzelius and the crystal-molecule epitomization theory of René Just Haüy, did not appear to be wholly compatible, and they became largely localized to the Germanic and to the French schools of chemistry, respectively.

Before Berzelius the two components of dualistic protochemical theory were not even-valued. On the basis of the body-spirit model, the alchemist sought, by pyrotechnical methods, to isolate and condense into tangible form the active spirits of composite substances. The later iatrochemists, recognizing a class of more potent noncondensible spirits, laid the foundations of eighteenth-century pneumatic chemistry, with its aim of characterizing the gases released from their gross integuments. For Antoine Laurent Lavoisier the active combustive principle of oxygen, the acid-generator, was central, the combustible sub-stances being passive recipients of the oxygen kernel released from its caloric shell, which provided the thermal effects of combustion. The active acids so generated condensed with bases as passive substrates to form the neutral salts. In the dualistic theory of Berzelius, the two constituent components of chemical substances were assigned equal status.[2] Oxygen, as the most electronegative of the elements, enjoyed no intrinsic primacy of place or function, apart from its practical convenience as a reference standard for chemical equivalents.

The stereochemical scope of dualistic theories proved to be limited, whether electrochemical, as with Berzelius, or caloric-based, as with John Dalton. Both theories agreed that like atoms repel one another, on account of either their like charge or the mutual repulsion of their caloric atmospheres. Electrochemical dualism had the advantage of explaining the combination of unlike atoms and of unlike radicals, through the coulombic attraction between the electropositive constituent and the electronegative constituent. Dalton's law of partial pressures, formulated in 1801, interpreting each gas as a vacuum to every other gas, appeared to imply that unlike atoms did not repel one another. Accordingly, the combination of unlike atoms was permitted, though not explained in terms of a positive affinity.

The forbidden combination of like atoms in both the caloric and the electrochemical theories ruled out the hypotheses of Amedeo Avogadro, formulated in 1811, and of André Marie Ampère, formulated in 1814, which required the elementary gases to be composed of even-numbered homoatomic molecules—diatomic in Avogadro's hypothesis and tetratomic in Ampère's.[3] Berzelius and his followers retained, however, a subsidiary implication of Gay-Lussac's law of gaseous combination by integral volumes, formulated in 1808, from which the Avogadro-Ampère hypothesis was derived in light of Dalton's atomic theory of 1808. The adoption of the view that the same volume of different elementary gases under given conditions of temperature and pressure contain the same number of atoms gave Berzelius the compositions H_2O for water and NH_3 for ammonia, as opposed to the Daltonian formulations HO and NH, respectively.

Without a rational basis for the valences of the elementary atoms, a theory of the internal structure of polyatomic molecules that would command a consensus was scarcely feasible. Only a partial consensus was achieved in fact after the Karlsruhe Congress of 1860, at which Stanislao Cannizzaro publicized his method, based upon Avogadro's hypothesis, for the determination of the valences and the atomic weights of the elements. The dualistic theory of Berzelius, with its prescription that even oleaginous organic substances are essentially composed of polar saline moieties, contributed little to the concept of valence. The most notable contribution of the dualistic school was that of Edward Frankland who, in 1852, found, from a comparison of organo-metallic derivatives with their corresponding inorganic compounds, that an electropositive organic radical took the place of one equivalent of oxygen, or other electronegative element, in completing the saturation capacity of the metal or metalloid element.[4]

Because there was not yet a theory of the internal structure of molecules, the external shape and overall form of polyatomic systems became the focus of attention, mainly from the French chemists of the early and middle nineteenth century. The French school had strong crystal-chemistry interests, initiated by Haüy, and either supported the hypothesis of Avogadro or, like Jean-Baptiste Dumas, considered the hypothesis significant enough to warrant an extended investigation.[5]

Haüy and mineralogists and crystallographers earlier in the eighteenth century found that, while a given substance may crystallize from a melt or a solution in different crystal habits dependent upon the particular conditions, cleavage reduces the various secondary forms to a common primary form with a regular polyhedral morphology. Were it feasible to continue the cleavage of the primary crystal form down to the ultimate molecular building blocks of the crystal, the molecular units would exhibit the same morphology as the primary crystalline polyhedron. As Haüy expressed the principle in 1809, the primary crystal and its constituent molecules are morphological "images of each other."[6]

That principle led Ampère in 1814 to his alternative version of Avogadro's hypothesis—that the molecules of the elementary gases were taken to be tetratomic, from the consideration that the building unit of a crystal must be necessarily three-dimensional and that the simplest polyhedron, the tetrahedron, requires four atoms, one at each vertex. Ampère's pupil Marc Antoine Augustin Gaudin reverted in 1833 to the diatomic formulation of Avogadro and used the vapor densities determined by Dumas in 1826—implying such gas-phase molecular constitutions as P_4, S_6, Hg—to draw up a protoperiodic table and, subsequently, to construct models of organic molecules. Alexandre Edouard Baudrimont in 1833 similarly supported Avogadro's hypothesis and used the crystal-molecule epitomization principle to question Berzelian dualism. He argued that Mitscherlich's law of isomorphism, stated in 1819, showed that molecular shape does not depend upon the particular electrochemical character of the constituent atoms. Isomorphic replacement in a crystal is analogous to a substitution reaction.[7]

The analogy between isomorphic replacement in a crystal and the substitution of electropositive hydrogen by an electronegative halogen, equivalent for equivalent, in an organic molecule was emphasized by Dumas in 1840 and by his pupil Auguste Laurent.[8] With a mineralogist's training at the Ecole des Mines, Laurent interpreted his pioneer studies of aromatic substitution and addition reactions in terms of crystal-molecule analogies, taking isomorphous substitution in a crystal as a model for atom replacement in a molecule. A hydrocarbon molecule constitutes a fundamental nucleus with a polyhedral shape that is preserved in substitution reactions, in which one atom or group replaces another. The products are derived nuclei with similar properties, often forming crystals isomorphic with those of the substrate.

During Laurent's lifetime, from 1807 to 1853, the polyhedral molecular shapes could be nothing more than conjectural guides to laboratory action. "It is impossible to know this arrangement," he stated, "but we may nevertheless ascertain whether in any particular body it is the same as in some other body."[9] Only with a reasonably assured set of atomic valences did conjectural molecular models of polygonal form—proposed by Kekulé in 1865—and of polyhedral form—proposed by Joseph Achille Le Bel and Jacobus H. van't Hoff in 1874 and by Alfred Werner from 1893 to 1914—give firmer and more detailed testable expectations as to the number and isomeric type of derived nuclei produced in substitution reactions of a given aromatic nucleus, aliphatic nucleus, or metal-coordination sphere.

Meanwhile, the investigation of molecular morphology by Louis Pasteur, based upon the crystal-molecule epitomization model, gave rise to the wholly novel conception of dissymmetry. Pasteur studied chemical crystallography with Laurent and, more particularly, with Gabriel Delafosse, one of Haüy's pupils. Delafosse in 1843 distinguished between chemical molecules and the polyhedral building units of crystals, the integrant molecules of Haüy, suggesting that each of the latter was made up of an ordered array of chemical molecules. Crystal properties additional to morphology—notably pyroelectricity and optical activity—reflected the structure and properties of the constituent molecules through the reticulation of their array in the crystal lattice.[10]

Haüy had recognized that there are morphological forms of the quartz crystal, distinguished by minor enantiomorphous facets that reduce the crystal symmetry from hexagonal to trigonal, but he regarded the facets as secondary, terming them *plagihedral*. The single primary form of quartz, Haüy supposed, was a hexagonal prism. Delafosse had regarded the enantiomorphous hemihedral facets as a primary property of the quartz crystal, reflecting a corresponding property of the constituent molecule, ever since John W. F. Herschel in 1822 had shown that quartz crystals of the two morphological sets, with a left- and a right-handed sequence of hemihedral facets, are optically levorotatory and dextrorotatory, respectively, to plane-polarized light (Fig. 1). Herschel observed that the hemihedral quartz facets "are produced by the same cause which determines the displacement of the plane of polarization of a ray traversing the crystal parallel to its axis."[11]

The particular stimulus leading Pasteur to his concept of dis-

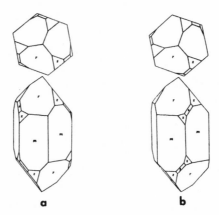

Fig. 1. Morphological forms of *a* left-handed (levorotatory) quartz crystals, and *b* right-handed (dextrorotatory) quartz crystals viewed in a direction parallel (upper view) and perpendicular (lower view) to the trigonal optic axis.

symmetry was the report by Eilhard Mitscherlich in 1844 of an earlier study connecting isomerism with isomorphism. The report stated that the sodium ammonium salt of (+)-tartaric acid and that of inactive racemic acid (paratartaric acid) are isomorphous and apparently identical in all physical and chemical respects, save optical activity. Working in Paris on his doctorate, Pasteur was astonished by the report, for all of the salts of (+)-tartaric acid that he had studied formed crystals with hemihedral facets. Repeating Mitscherlich's study of sodium ammonium racemate, Pasteur in 1848 found that two sets of hemihedral crystals are produced. Only one set proved to be truly isomorphous with crystals of sodium ammonium (+)-tartrate, with which it was identical in all properties, including specific optical rotation. The other set possessed, in addition to the enantiomorphous crystal facets, a specific optical rotation of opposite sign in aqueous solution. Following Haüy and, more particularly, Delafosse, Pasteur proposed that the spatial shapes of the individual (+)- and (−)-tartaric acid molecules are morphologically dissymmetric, related as nonsuperposable mirror-image forms, like the macroscopic crystals of the corresponding sodium ammonium salts (Fig. 2).[12]

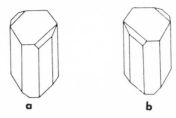

Fig. 2. Morphological forms of *a* dextrorotatory crystals and *b* levorotatory crystals of sodium ammonium tartrate.

Subsequently, while at Strasbourg from 1849 to 1854, Pasteur discovered the general method of diastereomer formation for the optical resolution of racemic substances, showing that a number of alkaloids form readily separable (+)- and (−)-tartrate salts with the racemic acid. He came to view dissymmetric molecules and hemihedral crystals as the product of universal dissymmetric forces, exemplified by the polar fields of electricity and magnetism, or composite rotations, such as those of the solar system. Following the discovery by Michael Faraday in 1846 of magnetically induced optical rotation in isotropic transparent media, Pasteur grew crystals with a normal holohedral habit in a magnetic field with the object of inducing dissymmetric hemihedral crystal forms. Not discouraged by the negative results, he attempted, while at Lille, from 1854 to 1857, to modify the optical activity of natural products by rotating the plants producing them with a large clockwork mechanism or, by means of a heliostat, presenting to the plants the appearance of the sun rising in the west and setting in the east. These experiments and their negative outcome were reported by Pasteur only in his later years, in the course of restating his belief in a cosmological dissymmetry. He argued that the solar, sidereal system is dissymmetric, being nonsuperposable upon its mirror-image.[13]

Back in Paris, Pasteur in 1860 indicated in his lectures on molecular dissymmetry that optical activity provides a provisional demarcation criterion between laboratory synthesis and the biochemical synthesis of living organisms. The latter had become his primary concern at Lille, where he had found that *Penicillium glaucum* grown on aqueous ammonium racemate con-

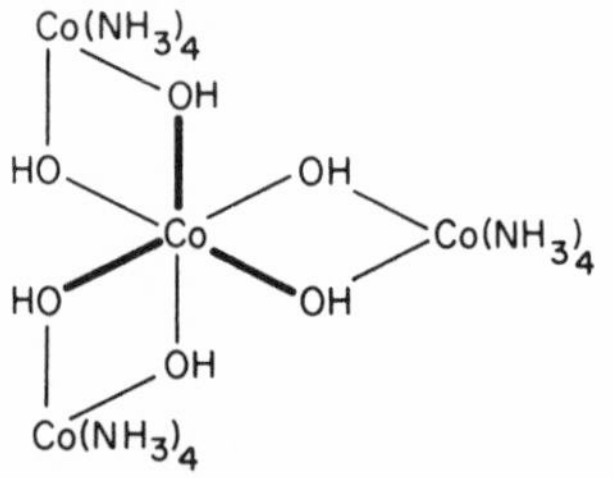

Fig. 3. The chiral metal coordination compound, containing no carbon, synthesized and optically resolved by Werner in 1914.

taining a small quantity of phosphate preferentially uses the (+)-tartrate as a carbon source, leaving the (−)-isomer. The distinction between the chemistry of the laboratory and that of living organisms, based upon optical activity, provided the justification for a vitalist school of organic stereochemistry, exemplified by Francis R. Japp at Aberdeen. That school, by postulating a chiral primacy for the carbon atom, provoked Werner to a synthetic *tour de force*, in the preparation of an optically active tetranuclear cobalt(III) coordination compound containing no carbon (Fig. 3).[14]

Pasteur himself found the vitalist school unattractive, and he was equally unconcerned with the contemporary developments of mainstream organic stereochemistry—first, the two-dimensional aromatic structural theory of Kekulé, introduced in 1865, and then the aliphatic chemistry in space of van't Hoff and Le Bel, introduced in 1874. Commenting in 1884 on the implications of his 1860 demarcation criterion, Pasteur observed, "Not only have I refrained from posing as absolute the existence of a barrier between the products of the laboratory and those of life, but I was the first to prove that it was merely an artificial barrier, and I indicated the general procedure necessary to remove it, by recourse to those forces of dissymmetry never before employed in the laboratory."[15]

Something of the iatrochemical tradition persisted in the viewpoint of Pasteur, reflected in his belief in the chemical autonomy of fermentation and disease, which were mediated by natural forces that, though as yet uncharacterized, could not be denied. The spontaneous separation of the optical isomers of sodium ammonium racemate upon crystallization might well be due, Pasteur supposed, to dissymmetric influences in the glass of the crystallizing dish. Once the dissymmetric forces of nature were understood and controlled, their use would lead to remarkable developments. "Success in this venture would give a new world of substances and reactions, and probably also of organic trans- formations. What can be said of the development of plant and animal species if it becomes possible to replace cellulose, al- bumin, and their analogues in the living cell by their optical antipodes?"[16]

Forces of dissymmetry, in addition to those considered by Pas- teur, were proposed by Le Bel in 1874 in his pioneer work on the connection between optical activity in the fluid phase and enan- tiomorphous three-dimensional organic structures. The use of left- or right-handed circularly polarized radiation in the pho- tochemical reactions of racemic substances, or of chiral cata- lysts in the corresponding thermal reactions, was expected to yield an excess of one particular enantiomeric product. Some twenty years later, A. Cotton in Paris discovered the isotropic circular dichroism of chiral molecules—the differential absorp- tion of left- and right-handed circularly polarized radiation— which is of equal magnitude but of opposite sign, for the two en- antiomers. Following his discovery, Cotton investigated the chi- ral photodiscrimination suggested by Le Bel, employing solu- tions of the coordination compound formed by copper(II) with racemic tartaric acid. No optical activity was detected by Cotton in the photoproducts, and twenty years passed before the first unambiguous photoresolutions were achieved.[17]

The second of the dissymmetric influences on reactivity pro- posed by Le Bel, the use of chiral catalysts, was implicit in the discovery of diastereomers by Pasteur in 1853, when the con- clusion was drawn that "the absolute identity of the physical and chemical properties of left and right non-superposable sub- stances ceases to exist in the presence of another active sub- stance." The introduction of a second asymmetric carbon center into an enantiomer gives two diastereomeric products that differ not only in chemical and physical properties but also in reaction yield. "There are equal chances for mirror images only," Emil Fischer observed in 1894. "Once a molecule is asymmetric, its extension proceeds also in an asymmetric sense. This concept completely eliminates the difference between natural and artificial synthesis. The advance of science has removed the last chemical hiding place for the once so highly esteemed *vis vitalis*."[18]

The expectations of van't Hoff as to the number and types of stereoisomers resulting from multiple chiral centers were both tested and used as a guide by Fischer in his investigation of the sugar series.[19] An aldohexose with four inequivalent chiral centers has sixteen stereoisomers; but the four asymmetric carbon atoms become two pairs of equivalent centers in the corresponding dicarboxylic acid, wherein the stereoisomerism is reduced to four enantiomeric pairs and two *meso* forms. The change from inequivalent to equivalent chiral centers, and the ascent and descent of the sugar series, served alike to support van't Hoff's guidelines and to correlate the configurations of the sugars.

From Fischer's work on the sugars arose the general problem of correlating the positive or negative sign of the optical rotation of an enantiomer—usually measured at the wavelength of the yellow sodium flame—with a particular three-dimensional model of the molecule or with the corresponding nonsuperposable mirror-image model. Fischer discovered that (+)-glucose, (+)-mannose, and (−)-fructose all formed the same osazone with phenylhydrazine, so that the three sugars must be stereochemically related, with the same configuration at the 3-, 4-, and 5-positions of the hexose chain (Fig. 4). After determining the relative configuration of the three positions by optically monitored chemical correlations, Fischer in 1891 ascribed an arbitrary absolute stereochemistry, then designated *d*, but now labeled the D-configuration, to the asymmetric carbon atom at the 5-position of each hexose studied. The hexose 5-position provides a representation of the unique chiral center in the then-unknown parent triose enantiomer, D-(+)-glyceraldehyde, from which all

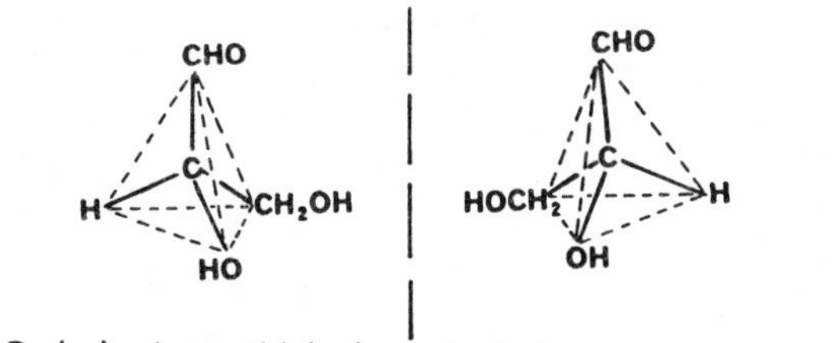

D-(+)-glyceraldehyde   L-(-)-glyceraldehyde

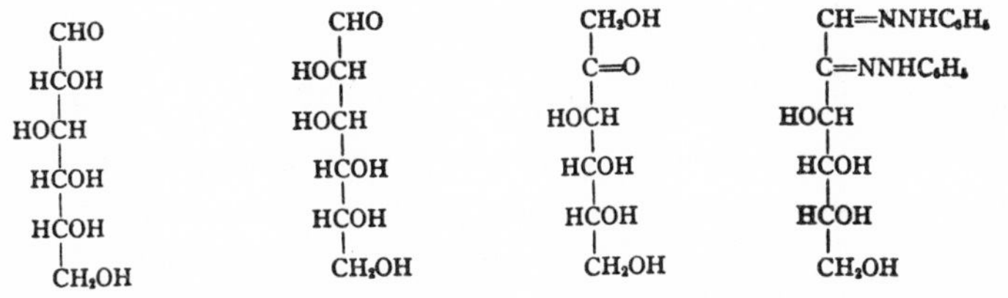

D-(+)-glucose   D-(+)-mannose   D-(-)-fructose   D-glucosazone

Fig. 4. The standard mirror-image structures of the Fischer-Rosanoff convention, D-(+)-glyceraldehyde and L-(-)-glyceralde-hyde, and the stereochemical relationship among the three sugars D-(+)-glucose, D-(+)-mannose, and D-(-)-fructose, and a common derivative with phenylhydrazine, D-glucosazone. In a conventional Fischer formula, the most oxidized carbon atom is placed uppermost, and in the D-hexose series, the configuration of the carbon atom at the 5-position correlates with that of the unique chiral center in D-(+)-glyceraldehyde. Conventionally, in a Fischer projection formula the vertical bonds from a carbon atom are directed to the rear, and the horizontal bonds have a forward orientation.

eight of the aldohexose diastereomers of the D-series were subsequently derived. The configurational convention of Fischer was reformed in 1906 by M. A. Rosanoff, who adopted D-(+)-glyceraldehyde and its L-(-)-enantiomer as the standard reference substances for the D- and the L-series of chiral molecules, respectively.[20]

By the time of the Rosanoff reform, it had become apparent that most naturally occurring sugars belong to the D-series and that all biologically important structural and functional α-amino acids belong to the L-series. The two biomolecular series were subsequently connected chemically by the conversion of D-glucosamine to L-alanine. A general explanation for the normal adoption of only one of the two series of sugars and amino acids by the biochemistry of living organisms, but not for the particular choice of the D-sugars and the L-amino acids, was provided by Fischer's "key and lock" hypothesis of enantioselective reactivity.[21] Compared to the racemic chemistry of the laboratory, the homochiral biochemistry is more economic and efficient, becoming thereby naturally selected.

A general optical solution to the problem of absolute stereochemical configuration was proposed by P. Drude in 1893, but his model was not successfully articulated until the 1960s, following further classical developments in the theory of optical activity and its reinterpretation in terms of quantum mechanics. Meanwhile, J. M. Bijvoet had developed an X-ray crystal-diffraction method for absolute stereochemical configuration, dependent upon the anomalous scattering of X-rays at near-absorption wavelengths. Application of this method showed in 1951 that Fischer and Rosanoff had fortunately, if fortuitously, made the correct configurational choice for their reference chirality.[22]

Following the morphological analogy between the primitive form of a crystal and the shape of the constituent molecular building units, Le Bel argued in 1890 that, if the valences of the carbon atom are directed to the apexes of a regular tetrahedron, a compound containing four identical groups bonded to a carbon atom should form isotropic crystals with cubic symmetry.[23] However, crystals of carbon tetrabromide and of the corresponding iodide proved to be birefringent; and so the crystals could not be exactly cubic, since they were not isotropic, though they appeared to be nearly cubic. Consequently, the four valences of a carbon atom could not be directed to the apexes of an exactly regular tetrahedron, though the distortions might well be small.

It also followed that the atoms of ethene could not be exactly coplanar and that substituted ethene derivatives should be resolvable into optical isomers. Accordingly, Le Bel attempted the

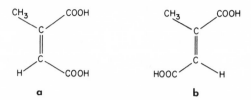

Fig. 5. The ethene derivatives that Le Bel in 1892 attempted to resolve optically: *a* citraconic acid and *b* mesaconic acid.

optical resolution of citraconic and mesaconic acids by Pasteur's third method, the use of microorganisms to metabolize one enantiomer of a racemic mixture (Fig. 5). Le Bel reported that optically active products were formed by the cultures, but subsequently he found that the products were not the original unsaturated acids. From the culture containing citraconic acid, Le Bel isolated (−)-methylmalic acid, formed by the addition of the elements of water under the enzymatic control of the microorganism to the carbon-carbon double bond of the substrate.[24]

By the 1890s a crystal could no longer be considered an assembly of congruent space-filling polyhedral units, each unit being a molecule or an ordered group of a small number of molecules. Through Auguste Bravais' work in 1850, the crystal came to be regarded as a reticular array of lattice points, symmetrically related by translational repetition. The lattice points represented the centers of gravity of the molecular units, but there was no necessary relation between the arrangement of the atoms in the molecule and the reticulation of the crystal lattice. The fourteen Bravais lattices accounted for seven of the thirty-two crystal point group classes, based upon the external crystal morphology first identified by J. F. C. Hessel and rediscovered by Bravais. Only the translational equivalence of the lattice points was taken into account by Bravais. In 1879 Leonhard Sohncke introduced the screw axis and the glide plane as lattice symmetry elements, extending the number of possible spatial arrangements of lattice points to sixty-five. Finally, E. S. Fedorov, A. Schoenflies, and W. Barlow separately, in the early 1890s, added the rotation-inversion lattice symmetry operation, completing the set of 230 space groups.[25] By the end of the century, the principal connection remaining between crystal morphology and molecular shape was the concept upon which the early work of Pasteur depended, namely, that an enantiomer cannot crystallize in a holohedral class with a space group containing secondary symmetry elements, though hemihedral facets are not necessarily observed in a crystal of a single enantiomer with a space group containing only primary symmetry elements.

At a meeting of the Chemical Society of France in 1924, celebrating the fiftieth anniversary of the discovery of the tetrahedral orientation of the valences of the carbon atom, Le Bel pointed out that the views on atomic structure that had developed over the intervening period implied that the chemical atom is intrinsically dissymmetric. According to the Bohr theory then prevailing, the electrons of an atom circulate around the nucleus like the planets around the sun. In some cases the circulation is clockwise and in other cases counterclockwise, so that a collection of atoms constitute a racemic mixture in most cases. Thus both sodium chlorate and strontium formate crystallize to give an equal number of levorotatory and dextrorotatory crystals. However, potassium silicotungstate and the corresponding molybdate crystallize to give predominantly dextrorotatory crystals, as G. Wyrouboff had first shown in 1896 and H. Copaux had confirmed in 1910. Hence, Le Bel concluded, the majority of the atoms of silicon belong to one of the two enantiomorphous sets. Similarly, the atoms of carbon must be enantiomerically enriched, since some of the native petroleums, which Le Bel held to have an inorganic origin, from metal carbides, are optically active.[26]

Subsequently, the results of Wyrouboff and Copaux on the predominant or sole formation of dextrorotatory crystals from aqueous solutions of potassium silicotungstate, wherein the complex is optically labile and equilibrated, were ascribed to trace impurities of alkaloids that had been used in concurrent optical resolutions in the respective laboratories. Frederic Stanley Kipping and William J. Pope had found in 1898 that the ratio of (+)-crystals to (−)-crystals of sodium chlorate, though 1:1

from aqueous solution, becomes 1:2 from aqueous glucose solution containing 20 percent of the sugar by weight. Surveys of the natural abundance of left- and right-handed quartz crystals have revealed a systematic minor excess abundance of (−)-quartz in all the localities sampled. In a total of 16,807 crystals, an enantiomeric excess of 1 percent of (−)-quartz has been documented.[27]

Universal dissymmetry ultimately emerged in a field of study nonexistent in Pasteur's day and just beginning to develop during the career of Le Bel. A group of anomalies in elementary particle physics led to the conclusion in 1956 that parity, or space-inversion equivalence, is not conserved in the weak nuclear interaction governing the β-decay of radionuclides. The observed form of the asymmetric β-decay of radionuclides showed the electron to be intrinsically left-handed, with antiparallel momentum and spin vectors, and the corresponding antiparticle, the positron, to be right-handed, with parallel momentum and spin vectors (Fig. 6).[28]

Chemists in the tradition of dissymmetry of Pasteur and Le Bel saw in the nonconservation of parity in the weak interaction the fundamental origin of the one-handed chirality of natural products, but the initial connections proved to be tenuous. It was not until the electromagnetic interaction, which mediates the binding energy of the valence electrons of atoms in molecular systems, was combined with the weak interaction that a chiral field governing the static ground-state properties of atoms and molecules appeared. One consequence of unification is universal optical activity in free atoms and nonchiral molecules as well as enantiomers. Such optical activity, which is very small, has been observed in vapors of heavy metal atoms.

Another consequence of the unification is the expectation of a small binding-energy difference between enantiomeric molecules. Calculations indicate that natural L-alanine is slightly more stable than its unnatural enantiomer, D-alanine, and that the naturally occurring L-polypeptides are correspondingly stabilized relative to their mirror-image D-forms. In both cases the difference is very small, corresponding to an enantiomeric excess of the L-amino acid or the L-polypeptide of a million mole-

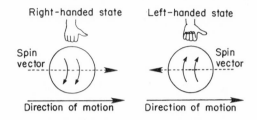

Fig. 6. The right-handed and the left-handed chirality of a particle, dependent upon the respective parallel or antiparallel relation between the linear momentum vector and the axial vector of
the angular momentum.

Stephen F. Mason, "Biomolecular Handedness," *Chemistry in Britain*, 21 (1985), 545.
Reproduced by permission.

cules in Avogadro's number, 6×10^{23}, for a racemic mixture in
thermodynamic equilibrium at ordinary temperatures. Although
small, the energy difference between the L-series and the D-
series of α-amino acids and polypeptides is adequate to explain,
on the basis of catastrophe-theory chemical kinetics, the transi-
tion from a racemic, or heterochiral, geochemistry to a single-
series, or homochiral, biochemistry in terrestrial evolution.
Moreover, the particular biomolecular homochirality found
by Fischer, specifically the L-amino acids and the D-sugars
chemically related to them, is rationalized. In a distant world of
antimatter, where the role of the electron is taken over by the
positron, and each nucleon is replaced by the corresponding
antinucleon, the converse biomolecular homochirality is expected to obtain, with D-amino acids and L-sugars.[29]

NOTES

1. G. P. Conger, *Theories of Macrocosms and Microcosms in the History of Philosophy* (New York, 1922); J. Needham, *Science and Civilisation in China* (Cambridge, U.K., 1976), Vol. 5, Pt. 3, pp. 143–45; G. E. R. Lloyd, *Polarity and
   Analogy: Two Types of Argumentation in Early Greek Thought* (Cambridge,
   U.K., 1966), 84.
2. E. M. Melhado, *Jacob Berzelius: The Emergence of His Chemical System*
   (Madison, Wis., 1981), 24.
3. J. H. Brooke, "Avogadro's Hypothesis and Its Fate: A Case-Study in the Failure
   of Case-Studies," *History of Science*, 19 (1981), 235–73; N. Fisher, "Avogadro,
   the Chemists and Historians of Science," *History of Science*, 20 (1982),
   82–102, 212–31.

4. E. Frankland, "On a New Series of Organic Bodies Containing Metals," *Philosophical Transactions of the Royal Society of London*, 142 (1852), 417–44.

5. S. C. Kapoor, "Dumas and Organic Classification," *Ambix*, 16 (1969), 1–65.

6. J. G. Burke, *Origins of the Science of Crystals* (Berkeley, Calif., 1966); S. H. Mauskopf, "Crystals and Compounds: Molecular Structure and Composition in Nineteenth-Century French Science," *Transactions of the American Philosophical Society*, 66, Pt. 3 (1976), 1–82; R. J. Haüy, *Tableaux comparatif des resultats de la crystallographie et de l'analyse chimique relativement à la classification des mineraux* (Paris, 1809), 17.

7. A. M. Ampere, "Lettre de M. Ampere à M. Le Conte Berthollet, sur la determination des proportions dans lesquelles les corps se combinent d'apres le nombre et la disposition respective des molécules dont leurs particules intégrantes sont composées," *Annales de Chimie*, 90 (1814), 43–86, Eng. trans. by S. H. Mauskopf, *Isis*, 60 (1969), 61–74; J. A. Miller, "Gaudin and Early Nineteenth Century Stereochemistry," in O. Bertrand Ramsey (ed.), *Van't Hoff-Le Bel Centennial* (Washington, D.C., 1975), 1–17; Brooke, "Avogadro's Hypothesis and Its Fate"; Fisher, "Avogadro, the Chemists and Historians of Science"; Mauskopf, "Crystals and Compounds."

8. S. C. Kapoor, "The Origin of Laurent's Organic Classification," *Isis*, 60 (1969), 477–527; N. W. Fisher, "Organic Classification before Kekulé," *Ambix*, 20 (1973), 106–31, 209–33.

9. A. Laurent, *Chemical Method*, trans. W. Odling (London, 1855), 324.

10. S. H. Mauskopf, "Crystals and Compounds."

11. *Ibid.*; J. W. F. Herschel, "On the Rotation Impressed by Plates of Rock Crystal on the Planes of Polarization of the Rays of Light as Connected with Certain Peculiarities in its Crystallization," *Transactions of the Cambridge Philosophical Society*, 1 (1822), 43–50.

12. J. B. Biot, "Communication d'une note de M. Mitscherlich," *Comptes Rendus Hebdomadaires des Séances de l'Academie des Sciences* (Paris), 19 (1844), 719; L. Pasteur, "Mémoire sur la relation qui peut exister entre la forme crystalline et la composition chimique, et sur la cause de la polarisation rotatoire," *Comptes Rendus Hebdomadaires des Séances de l'Academie des Sciences* (Paris), 26 (1848), 535–38; Pasteur Vallery-Radot (ed.), *Oeuvres de Pasteur* (7 vols.; Paris, 1922), Vol. 1, pp. 61–64.

13. L. Pasteur, "Nouvelles Recherches sur les relations qui peuvent exister entre la forme crystalline, la composition chimique, et le phénomène rotatoire moléculaire," *Annales de Chimie et de Physique*, 3rd ser., 38 (1853), 437–83; Vallery-Radot (ed.), *Oeuvres de Pasteur*, Vol. 1, pp. 203–41; M. Faraday, "On the Magnetisation of Light and the Illumination of Magnetic Lines of Force," *Philosophical Transactions of the Royal Society of London* (1846), 1; *Philosophical Magazine*, 28 (1846), 294; L. Pasteur, "La Dissymétrie moléculaire," *Revue Scientifique*, 7 (1884), 2; Vallery-Radot (ed.), *Oeuvres de Pasteur*, Vol. 1, pp. 369–80.

14. L. Pasteur, *Recherches sur la dissymétrie moléculaire* (1860); Vallery-Radot (ed.), *Oeuvres de Pasteur*, Vol. 1, pp. 314–44, Eng. trans., *Alembic Club Reprint*, No. 14 (Edinburgh, 1948); L. Pasteur, "Mémoire sur la fermentation de l'acide tartarique," *Comptes Rendus Hebdomadaires des Séances de l'Academie des Sciences* (Paris), 46 (1858), 615–18; Vallery-Radot (ed.), *Oeuvres de Pasteur*, Vol. 2, pp. 25–28; F. R. Japp, "Stereochemistry and Vitalism," *Na-

ture, 58 (1898), 452–60; A. Werner, "Zur Kenntnis des asymmetrischen Kobaltatoms XII. Über optische Aktivität bei kohlenstofffreien Verbindungen," *Berichte der Deutschen Chemischen Gesellschaft*, 47 (1914), 3087–94.

15. L. Pasteur, "Aux remarques de MM. Wyrouboff et Jungfleisch sur 'La Dissymétrie moléculaire,'" *Bulletin de la Société Chimique de France*, 41 (1884), 215–20; Vallery-Radot (ed.), *Oeuvres de Pasteur*, Vol. 1, p. 385.

16. Pasteur, "La Dissymétrie moléculaire," and "Observations sur les forces dissymétriques," *Comptes Rendus Hebdomadaires des Séances de l'Academie des Sciences* (Paris), 78 (1874), 1515–18; Vallery-Radot (ed.), *Oeuvres de Pasteur*, Vol. 1, p. 360.

17. J. A. Le Bel, "Sur les relations qui existent entre les formules atomiques des corps organiques, et le pouvoir rotatoire de leurs dissolutions," *Bulletin de la Société Chimique de France*, 22 (1874), 337–47; A. Cotton, "Recherches sur l'absorption et la dispersion de la lumière par les milieux doués du pouvoir rotatoire," *Annales de Chimie et de Physique*, 7th ser., 8 (1896), 347–437, "Absorption inégle des rayons circulaires droit et gauche dans certains corps actifs," *Comptes Rendus Hebdomadaires des Séances de l'Academie des Sciences* (Paris), 120 (1895), 989–91, "The Application of Ciruclar Dichroism to the Synthesis of Active Compounds," *Transactions of the Faraday Society*, 26 (1930), 377–83, and "Sur le dédoublement des corps inactifs par compensation obtenus par synthèse chimique," *Journal de Chimie Physique*, 7 (1909), 81–96; W. Kuhn, "The Physical Significance of Optical Rotatory Power," *Transactions of the Faraday Society*, 26 (1930), 293–308; W. Kuhn and E. Braun, "Photochemische Erzeugung optisch aktiver Stoffe," *Naturwissenschaften*, 17 (1929), 227–28.

18. Le Bel, "Sur les relations qui existent entre les formules atomiques"; Pasteur, "Nouvelles Recherches sur les relations"; E. Fischer, "Synthesen in der Zuckergruppe," *Berichte der Deutschen Chemischen Gesellschaft*, 27 (1894), 3189–232.

19. J. H. van't Hoff, "Sur les formules de structure dans l'espace," *Archives Néerlandaises des Sciences Exactes et Naturelles*, 9 (1874), 445–54.

20. E. Fischer, "Ueber die Configuration des Traubenzuckers und seiner Isomerism," *Berichte der Deutschen Chemischen Gesellschaft*, 24 (1891), 2683–87; M. A. Rosanoff, "On Fischer's Classification of Stereo-Isomers," *Journal of the American Chemical Society*, 28 (1906), 114–21.

21. M. L. Wolfrom, R. U. Lumieux, and S. M. Olin, "Configurational Correlation of L-(levo)-Glyceraldehyde with Natural (dextro)-Alanine by a Direct Chemical Method," *Journal of the American Chemical Society*, 71 (1949), 2870–73; Fischer, "Synthesen in der Zuckergruppe."

22. P. Drude, *The Theory of Optics*, trans. C. R. Mann and R. A. Millikan (New York, 1959), 400–17; S. F. Mason, *Molecular Optical Activity and the Chiral Discriminations* (Cambridge, U.K., 1982), 18–137, 240–47; J. M. Bijvoet, A. F. Peerdeman, and A. J. van Bommel, "Determination of the Absolute Configuration of Optically Active Compounds by Means of X-Rays," *Nature*, 168 (1951), 271–72.

23. J. A. Le Bel, "Sur les conditions d'équilibre des composés saturés du carbone," *Bulletin de la Société Chimique de France*, 3rd ser., 3 (1890), 788.

24. J. A. Le Bel, "Sur le changement de signe du pouvoir rotatoire," *Bulletin de la*

Société Chimique de France, 3rd ser., 7 (1892), 613, and "Essais de dédouble- ment de corps non saturés," *Bulletin de la Société Chimique de France*, 3rd ser., 11 (1894), 258, 292.

25. A. Bravais, "On the Systems Formed by Points Regularly Distributed on a Plane or in Space," *Journal de l'École Polytechnique* (Paris), 19 (1850), 1; 20 (1851), 101 (Eng. trans. A. J. Shaler, American Crystallographic Association Monograph No. 4, 1969); J. F. C. Hessel, *Krystallometrie* (Leipzig, 1831), rpr. in E. Hess (ed.), *Ostwald's Klassider der Exakten Wissenschaften*, Nos. 88 and 89 (Leipzig, 1897); L. Sohncke, *Die Entwicklung einer Theorie der Kristallstruktur* (Leipzig, 1879); E. S. Fedorov, "Symmetry of Crystals," *Transactions of the Mineralogical Society*, 28 (1891), 1; Eng. trans. D. Harker and K. Harker (American Crystallographic Association Monograph 7, 1971); A. Schoenflies, *Kristallsysteme und Kristallstruktur* (Leipzig, 1891); W. Barlow, "A Mechanical Cause of Homogeneity of Structure and Symmetry," *Proceedings of the Royal Dublin Society*, n.s., 8 (1897), 527–690.

26. J. A. Le Bel, "Discours à l'occasion du cinquantenaire de la théorie du car- bone asymétrique," *Bulletin de la Société Chimique de France*, 4th ser., 37 (1925), 353; M. Delepine, *Vie et Oeuvres de Joseph Achille Le Bel* (Paris, 1949), 17; G. Wyrouboff, "Recherches sur les silicotungstates," *Bulletin de la Société de Mineralogie de France*, 19 (1896), 219–354; H. Copaux, "Sur l'inégalité de propriétés des deux variétés, droite et gauche, du silicotungstate de po- tassium," *Bulletin de la Société de Mineralogie de France*, 33 (1910), 167–73.

27. A. Amariglio, H. Amariglio, and X. Duval, "La Synthèse asymétrique," *An- nales de Chimie* (Paris), 3 (1968), 5–25; F. S. Kipping and W. J. Pope, "Enan- tiomorphism," *Journal of the Chemical Society*, 73 (1898), 606–17; C. Palache, H. Berman, and C. Frondel, *Dana's System of Mineralogy* (7th ed., 3 vols.; New York, 1962), Vol. 3, p. 16.

28. Mason, *Molecular Optical Activity*.

29. S. F. Mason, "Origins of Biomolecular Handedness," *Nature*, 311 (1984), 19–23, and *Molecular Optical Activity*.

The Early History and Development of Conformational Analysis

In 1975 the Nobel Prize in chemistry was awarded jointly to John W. Cornforth and Vladimir Prelog for their "researches on the stereochemistry of enzyme-catalyzed reactions and the stereochemistry of organic molecules and reactions, respectively." This was the second Nobel Prize awarded within six years for contributions in the field of stereochemistry. The first was awarded in 1969, also jointly, to the English chemist Derek H. R. Barton and the Norwegian physical chemist Odd Hassel for "developing and applying the principles of conformation in chemistry."

The research of Prelog, Barton, and Hassel centered around an area of chemistry known as conformational analysis. The development of conformational analysis focused on the answers to two fundamental questions: 1) What is the shape of the cyclohexane ring? and 2) Is there a rotational barrier around the carbon-carbon single bond? An appreciation of the importance of the nature and use of molecular models and the importance of scientific communication also characterized the development of conformational analysis. The publication of two papers in 1950 by Barton and Prelog marked a shift from "static" stereochemistry to "dynamic" stereochemistry. Until the 1950s, molecular models had been generally used to illustrate static structural features of molecules. In tracing the development of conformational analysis, we see the use of models to illustrate the dynamic interconversions of molecular structures that allowed the chemist to demonstrate the relationship between a substance's structure and its chemical reactivity.

That the use of molecular models was not considered part of a serious approach in the study of structural problems even in the

1950s is suggested in James Watson's account of his and Francis Crick's discovery of the structure of DNA. Discussing the importance of Linus Pauling's proposal for the α-helical structure of protein, he said, "The α-helix had not been found by only staring at X-ray pictures; the essential trick, instead, was to ask which atoms like to sit next to each other. In place of pencil and paper, the main working tools were a set of molecular models superficially resembling the toys of preschool children. We could thus see no reason why we should not solve DNA in the same way. All we had to do was to construct a set of molecular models and begin to play—with luck the structure would be a helix." The use of molecular models by Watson and Crick proved crucial to their determination of the structure of DNA. Apparently, Rosalind Franklin, who had performed the critical X-ray studies that served as the basis of the Watson-Crick model of DNA, was among others who were not easily convinced of the value of models. Watson wrote, "Obviously affecting Rosy's transformation was her appreciation that our past hooting about model building represented a serious approach to science, not the easy resort of slackers who wanted to avoid the hard work necessitated by an honest scientific career."[1]

That the shape of the cyclohexane ring was still a viable question in the 1940s is apparent in the opening sentence of Barton's 1950 paper titled "The Conformation of the Steroid Nucleus": "In recent years it has become generally accepted that the chair conformation of cyclohexane is appreciably more stable than the boat." Barton was able to make this statement about the greater stability of the chair form because of calculations he had undertaken two years earlier. In the earlier paper he assumed that the relative stabilities of the chair and the boat forms of cyclohexane, as in the staggered and the eclipsed forms of ethane, would be determined by the sum of the attractive and the repulsive forces experienced by the nonbonded hydrogens. To measure the distances between the nonbonded atoms, Barton used specially constructed molecular models made from aluminum (Fig. 1).[2] These models were later sold commercially by the Wilkenson-Anderson Company in Chicago.

Barton has indicated that he wrote his 1950 paper after listen-

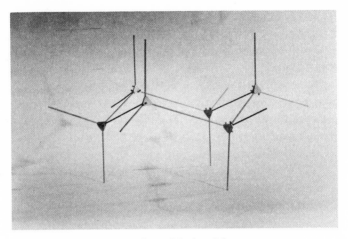

Fig. 1. Barton's model of cyclohexane.

ing to a lecture by Louis Fieser at Harvard University in the fall of 1949. Barton was able to provide the solutions to a number of stereochemical problems posed by Fieser because of his familiarity with the electron-diffraction studies undertaken in the 1930s and 1940s by the Norwegian physical chemist Hassel. Although Hassel's key paper was published in 1943, Barton became aware of Hassel's work only through an English review published in 1946. That there was no general agreement among chemists as to the shape of the cyclohexane ring is clear from the title, "The Cyclohexane Problem," and the opening line of Hassel's 1943 paper: "If one assumes that the valence angles of the carbon atom are equal to the 'tetrahedral' angle (109° 28'), or in any case not substantially different from this angle, then the possibility that the carbon atoms in cyclohexane form a coplanar six-membered ring is excluded."[3]

How was it possible for chemists in the 1940s to seriously consider a planar cyclohexane ring? The idea of a planar cyclohexane ring can be traced back to a proposal made in 1885 by Adolf Baeyer, one of the first chemists to adopt the stereochemical concepts proposed by Jacobus van't Hoff and Joseph A. Le Bel in 1874. Although the basic ideas proposed by van't Hoff and Le Bel were the same, van't Hoff was more influential in the subsequent development and acceptance of stereochemical concepts by chemists, in part because of his numerous books on stereochemistry and his effective use of perspective drawings and molecular models to illustrate the stereochemical concepts. Van't Hoff's 1874 pamphlet included a plate containing fifteen structural diagrams, five of which were perspective drawings (Fig. 2). His 1875 revised edition contained three plates with sixty diagrams and seventeen perspective drawings.[4] The tetrahedral orientation of the carbon valences is clearly illustrated in the enantiomeric pair included in the 1874 publication (VII and VIII in Fig. 2).

Since van't Hoff was uncertain that even these kinds of illus-

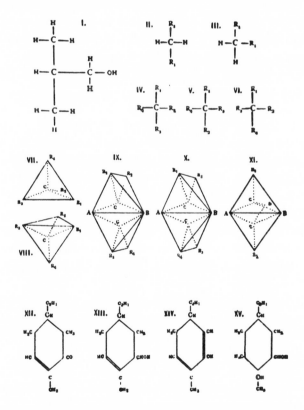

Fig. 2. Structural diagrams in van't Hoff's 1874 pamphlet.

trations would be understood by the readers of his 1875 publication, he constructed several sets of cardboard models that he sent to a number of prominent chemists such as August Kekulé, Johannes Wislicenus, Charles A. Wurtz, and Baeyer. Baeyer, however, apparently made no use of them in the subsequent development of his ideas on stereochemistry.

Unlike Le Bel, van't Hoff utilized the concept of the tetrahedral carbon to explain the existence of cis/trans isomers, which he designated "geometric" isomers (IX and X in Fig. 2). No experimental evidence supported the tetrahedral orientation of valences in acetylene (XI in Fig. 2). In 1875 van't Hoff extended the concept of cis/trans isomerism to include cyclic compounds (Fig. 3). Baeyer, however, was unaware of these ideas, relying on the information in the German edition of van't Hoff's work, published in 1877, which did not include a discussion of the cis/trans isomerism possible for cyclic compounds.[5] In the 1880s Baeyer published a series of papers concerning the structure of substituted cyclohexane compounds. To distinguish the cis/trans isomers, Baeyer used the terms *malenoid* and *fumaroid*.

In a paper written in 1885, Baeyer described the preparation and the properties of polyacetylenes. To account for the explosive instability of these compounds, he suggested that there was a strain or tension (*spannung*) in the triple bond. In an addendum to the paper, Baeyer examined the stabilities of cyclic compounds. At that time only the 6-membered ring was known, as found in benzene and cyclohexane derivatives. Attempts to prepare smaller and larger rings had not been successful. Baeyer based his strain theory on van't Hoff's concept of the tetrahedral carbon.

> The direction of these attractions can undergo a diversion which causes a strain which increases with the size of the diversion. The meaning of this statement can easily be explained if we start from the Kekulé spherical model and assume that the wires like elastic springs are movable in all directions. If now, the explanation that the direction of the wires is also assumed, a true picture is obtained. . . .
> If now, as can already be shown by the use of a model, an attempt is made to join a greater number of carbon atoms without force, that is, in the direction of the tetrahedral axes, or the wires of the models,

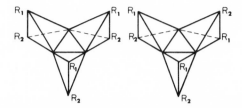

Fig. 3. van't Hoff's representation of cyclic cis/trans isomers.

the result is either a zig zag line or a ring of five atoms, which is entirely comprehensible since the angles of a regular pentagon, 108°, differ only slightly from the angle 109° 28' which the axes of attraction make with one another. When a larger or smaller ring is formed the wires must be bent, that is, there occurs a strain.[6]

Baeyer went on to calculate the angular distortion found in ethylene and in the 3- through 6-membered rings (Fig. 4). Today the reader of the above passage would have no difficulty visualizing the origin of the strain in most of these compounds. Most chemists have at one time or another used molecular models in which the bonds are made out of wires or springs that must be bent in order to form a double or a triple bond or a small-ring cyclic structure. It is also apparent, however, that when the models are used to construct the cyclochexane ring, there is no strain. The strain proposed by Baeyer requires a planar ring, which would not be observed when these conventional models are used.

How was it possible for Baeyer to have missed this observation? Some historians have speculated that perhaps he did not comment on this apparent inconsistency because he assumed that cyclohexane must be planar owing to its experimentally demonstrated structural relationship to benzene. If the hydrogen atoms in benzene are coplanar, is it unreasonable to think that the carbon atoms are also coplanar? And if so, might not this coplanarity be maintained when benzene is converted to cyclohexane?

Historians have overlooked one detail in the above description. Baeyer stated in 1885 that he used "Kekulé spherical models," first described by Kekulé in 1867. To account for the

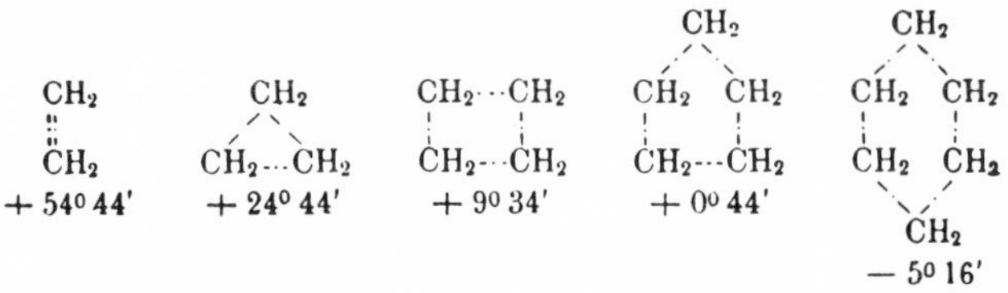

$$\begin{array}{ccccc}
& & & \text{CH}_2 & \text{CH}_2 \\
& & & \diagup\diagdown & \diagup\diagdown \\
\text{CH}_2 & \text{CH}_2 & \text{CH}_2\cdots\text{CH}_2 & \text{CH}_2 \quad \text{CH}_2 & \text{CH}_2 \quad \text{CH}_2 \\
\vdots & \diagup\diagdown & \vdots\quad\vdots & \vdots\quad\quad\vdots & \vdots\quad\quad\vdots \\
\text{CH}_2 & \text{CH}_2\cdots\text{CH}_2 & \text{CH}_2\cdots\text{CH}_2 & \text{CH}_2\cdots\text{CH}_2 & \text{CH}_2 \quad \text{CH}_2 \\
+54^{\circ}44' & +24^{\circ}44' & +9^{\circ}34' & +0^{\circ}44' & \diagdown\diagup \\
& & & & \text{CH}_2 \\
& & & & -5^{\circ}16'
\end{array}$$

Fig. 4. Baeyer's assessment of angular distortion in some
hydrocarbons.

limitations of the two-dimensional graphic formulas then in use,
Kekulé proposed that the carbon valences were tetrahedrally
oriented in space, some seven years before van't Hoff and Le Bel
made their proposals. In his lectures, Kekulé illustrated the
tetrahedral valence concept with molecular models. These
models were subsequently used by a number of Kekulé's students, including Baeyer, and were available commercially in the
1880s. Baeyer used these models in somewhat modified form as
the basis of his strain theory. William Henry Perkin, Jr., a student of Baeyer in Munich in the 1880s, provided the following
description of the models: "I remember that on two occasions
[Baeyer] invited me into his study and explained to me, with the
aid of models that had been especially made, his views on the
stability and ease of formation based on the tetrahedral configuration of the carbon atom, and these views gradually gave rise to
the 'Spannungs Theorie.'"[7]

The structure of cyclohexane constructed with Kekulé-Baeyer
models is shown in Figure 5. The strain in the bonds is proportional to the angles between the wires. In the case of cyclohexane, the wires point inward and there is a "negative" strain.
The major features of this model are that the carbon atoms are
coplanar, the tetrahedral angles of the valence wires are maintained, and the inward direction of the valence wires illustrates
the negative ring strain.

Since the Baeyer strain theory played such an important role
in the subsequent development of conformational analysis, it is
perhaps worthwhile to examine the basis of this theory in more
detail. One might first ask whether Baeyer thought there was
any experimental support for this structure of cyclohexane. In

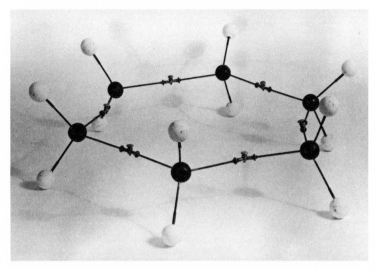

Fig. 5. Kekulé-Baeyer model of cyclohexane.

1890 he observed that the reduction of phthalic acid gave only the cis isomer of 1,2-cyclohexanedicarboxylic acid. He thought this stereoselectivity consistent with the centric formula of benzene that he had proposed earlier, which placed all six tetrahedral carbons and six hydrogens on one side of the ring (Fig. 6). If upon reduction the six hydrogens added on the opposite side of the ring, only the cis isomers should be formed. In addition, Baeyer had found that a cyclic anhydride could be prepared from both the *cis-* and *trans*-1,2-cyclohexanedicarboxylic acid. "In fumaric acid the direction of the valencies binding the carboxyls are exactly opposite and make an angle of 180° with each other, while in *fumaroid* hexahydrophthalic acid this angle is only a little more than 109°. . . . This is shown in the following figures [see Fig. 7], but these cannot give an accurate picture of the angles without reference to the model."[8]

To appreciate the latter point, the reader is asked to imagine the adjacent trans hydrogens—one above and one below the plane of the ring in Figure 5—replaced by carboxyl groups. Looking along the carbon-carbon bond connecting these carboxyl groups, one can see that the angle that might now be des-

Fig. 6. Kekulé-Baeyer model of Baeyer's centric formula for benzene.

$$CO_2H \quad H \qquad CO_2H \quad H$$

fumaroïde
Hexahydrophtalsäuer

Fumarsäuer

Fig. 7. Illustrations based on those in Baeyer's 1888 article about 1,2-cyclohexanedicarboxylic acids (hexahydrophthalic acids).

ignated as the dihedral angle approximates 109°. (The dihedral angle for the cis carboxyl groups would be about 0°.)

The number of chemists who may have used the Kekulé models in the nineteenth century and the early twentieth century to illustrate the strain theory is unknown. It is clear, however, that most chemists in this period did accept the coplanarity of the carbon atoms in cyclohexane.

In a paper titled "On the Geometrical Isomers of Hexamethylene Derivatives," published in 1890, Hermann Sachse attempted to explain the apparently facile chemical interconversion of the cis/trans isomers of polycarboxylic acids observed by Baeyer in his investigations. Sachse proposed that cyclohexane exists in one of two strain-free conformations (he used the term *configurations*): a "symmetrical," rigid chair form or an "unsymmetrical," flexible boat form. That Sachse had noted the greater flexibility of the boat conformation suggests that he used Kekulé models. In his paper, however, he provided instructions on how to prepare models from cardboard.

> The "normal" [chair] form . . . can be obtained if the six triangles in Figure [8] are covered with matching regular tetrahedra in such a way that the first, third and fifth tetrahedra are situated above the plane of the drawing. The system also possesses, among other things, the remarkable characteristic that the six tetrahedra assume a position as if they are resting on the six surfaces of an octahedron. [The lower illustration in Figure 9 was in a footnote about the method of construction.] The "non symmetric" form can be represented in the following way. Two (incomplete) octahedron models, obtained from 2 pieces shaped as illustrated in Fig. [9, top] in the way already given, are placed together in such a way that the places marked with the same letters meet at one point, the shaded triangles are covered with matching tetrahedra.[9]

Preparing models according to these instructions is not easy, and there is little evidence that any chemists in this period took the time to construct them. The poor reception of Sachse's ideas may be attributed not merely to the chemists' unwillingness to construct molecular models but even more to a suggestion by Sachse, capable of experimental verification. In brief, Sachse proposed that a chair-chair interconversion was responsible for the rapid chemical interconversion of the cis/trans isomers of substituted cyclohexanes. Thus *cis*-1,2-cyclohexanedicarboxylic acid, with diequatorial carboxyl groups, was converted to the trans isomer, with diaxial carboxyl groups. Sachse pointed out that a third isomer (the axial-equatorial form) was possible but was presently unknown. Since the rapid interconversion of the cis/trans isomers was considered to arise from the natural repulsion between the carboxyl groups, Sachse suggested that it should be possible to isolate the two chair conformations of monosubstituted cyclohexanes where these repulsions had been eliminated. Since subsequent experimental work did not confirm this prediction, a planar cyclohexane ring seemed more probable.

Sachse's ideas fell into virtual oblivion for some twenty-five years, until they were revived in a series of papers published by

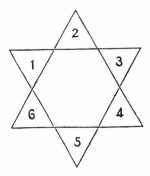

Fig. 8. Sachse's drawing for constructing a cyclohexane model.

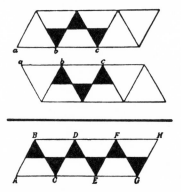

Fig. 9. Illustrations in Sachse's 1890 instructions for preparing cyclohexane models.

Ernst Wilhelm Max Mohr, a professor of chemistry at the University of Heidelberg. Mohr pointed out that Sachse's concept of the multiplanar cyclohexane ring could be used to explain the existence of a number of bicyclic and tricyclic compounds, such as camphor and adamantane, whose structures were difficult to visualize in terms of the planar rings proposed by Baeyer.[10] Mohr's papers had more appeal to chemists than Sachse's, since they included numerous perspective drawings apparently taken from Kekulé models (Fig. 10). In addition to pointing out that the cis/trans isomerization of disubstituted cyclohexanes did not arise from a chair-chair interconversion, Mohr also pointed out that such a conversion for the monosubstituted cyclohexanes was likely to be too rapid to allow for the isolation of the conformational isomers.

The existence of the multiplanar ring could be demonstrated in the search for cis/trans isomers in fused-ring compounds such as decalin, wherein the ring fusion prevents the chair-chair interconversion (Fig. 10). This prediction was confirmed experimentally in 1925 by Walter Hückel, who isolated the two diastereomers of a decalin derivative.[11]

In spite of this evidence, the concept of a planar cyclohexane ring persisted in the literature. Experimental work undertaken in the early part of the twentieth century provided ambiguous and often contradictory results regarding the existence of the multiplanar cyclohexane ring. For most chemists it was of no practical importance whether the ring was planar or not. The Finnish chemist Ossian Aschan represented a generally held view. "The fact that this view of the configuration of the hexamethylenes is not verified is probably to be explained on the grounds that Sachse's formulas represent only different phases

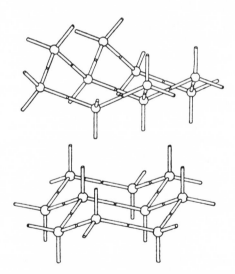

Fig. 10. Mohr's drawings of decalin isomers.

Ernst Mohr, "Die Baeyersche Spannungstheorie und die Struktur des Diamanten," *Journal für praktische Chemie*, 98 (1918), 321. Reproduced by permission.

of the movement within the molecules of the hexamethylenes."[12]

The planar ring received additional experimental support as early as 1915 from the influential founders of physical organic chemistry in England, Christopher Ingold and Jocelyn Thorpe. "It is, of course, assumed there is no distortion of the angle of hexagon caused by the attachment of groups to any one carbon atom of the cyclohexane ring." Fifteen years later Thorpe had not altered his views in spite of the fact that by then he and Ingold had provided the experimental evidence in favor of the multiplanar cyclohexane ring. "There is then considerable evidence in favor of a strained cyclohexane structure both from the chemical and physical side and none whatever in favor of the strainless multiplanar ring."[13]

The nonapplicability of the strain theory to large ring systems was demonstrated experimentally in 1926 by Leopold Ruzicka, who showed that the substances muskone and civetone were cyclic ketones containing fifteen and seventeen carbon atoms. Even though Ruzicka and his coworkers had devised syntheses of a number of these macrocyclic compounds, the 6-membered ring remained an apparent anomaly for Ruzicka at least as late as 1935. "The circumstances existing in the 6-ring are not the same as in the higher-membered rings, the stability of which can only be explained by the acceptance of Sachse's theory."[14]

The pervasiveness of the idea of the planar cyclohexane ring can be seen in the interpretations of the first X-ray crystallographic studies of cyclohexane derivatives. In 1926 Sterling B. Hendricks and Constant Bilicke at the California Institute of Technology published the results of their crystallographic analysis of the β-isomer of 1,2,3,4,5,6-hexabromocyclohexane. Although it seems clear from the figures in the article that the cyclohexane carbons are in a chair conformation (Fig. 11), Hendricks and Bilicke seemed unwilling to concede that cyclohexane might be multiplanar.

> Mohr's theory of "Strainless Rings" as applied to cyclohexane and its derivatives is not compatible with our conclusions in so far as the solid state is concerned. His three-dimensional formulae . . . have a center of symmetry in one case, but not a plane of symmetry. . . . Four carbon atoms of a particular cyclohexane ring are coplanar, the

1,4-carbon atoms being equidistantly placed above and below this plane. Such a representation would not be tenable on the basis of our conclusions. A promiscuous interconversion of molecular configurations between two widely separated forms as represented by Fig. [11] and as required by Mohr's theory, would probably not be possible.

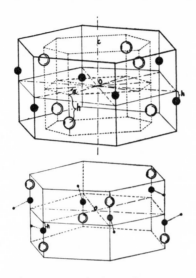

Arrangement of the carbon atoms (large black dots) and halogen atoms (white circles) in molecule of $C_6H_6X_6$. The small black dots represent hydrogen atoms.

Fig. 11. Figures used by Hendricks and Bilike in 1926 for the β-isomer of 1,2,3,4,5,6-hexabromocyclohexane.

Sterling B. Hendricks and Constant Bilicke, "The Space-Group and Molecular Symmetry of β-Benzene Hexabromide and Hexachloride," *Journal of the American Chemical Society*, 48 (1926), 3013. Reproduced by permission.

Although Roscoe Dickinson and Bilicke provided additional evidence of the multiplanar structure two years later, they seemed unwilling to support explicitly Mohr's theory of the multiplanar structure. "These parameter values lead to reasonable interatomic distances and are in agreement with a molecule containing a cyclohexane ring of 'tetrahedral' carbon atoms."[15]

The other question affecting the development of conformational analysis, namely, whether a rotational barrier existed around the carbon-carbon single bond, interested van't Hoff as early as 1875. He discussed and provided illustrations of the various "phases" of the molecule that would be produced as a result of rotations around a single bond.[16] Although van't Hoff recognized that there might be a "favored" phase of the molecule as a result of the different forces of repulsion experienced by the substituents, he concluded that for all practical purposes the rotation must be free around the bond, since otherwise an excessive number of isomers would be encountered. Most chemists in the nineteenth century and the early twentieth century adopted this view.

One scientist who did not adopt this view was Carl Bischoff, a professor of chemistry at Riga Polytechnic Institute. In 1891 Bischoff proposed that the staggered form of ethane should be more stable than the eclipsed form (II and I, respectively, in Fig. 12) because of the repulsions between the nonbonded hydrogens.[17] Since other substituents should exhibit a greater repulsive effect than hydrogen, Bischoff argued that the favored conformation of the molecule should be that which placed these substituents at a maximum distance from one another. That methyl-substituted succinic acids formed cyclic anhydrides more easily than did succinic acid itself was viewed by Bischoff as arising from the repulsions between the methyl groups, which produced a favored conformation in which the two carboxyl groups were closer to each other (Fig. 13) than in succinic acid, in which they were in an anti conformation (Fig. 14).

At one point Bischoff thought he had isolated an optically active form of one of the substituted succinic acids, a result that he interpreted as indicative of the thermal stability of the favored conformation. Subsequent experimental studies, however, did not support this observation, and Bischoff's speculations fell into oblivion.

Not until the 1920s did quantitative work provide any insights into the question of a rotational barrier. These studies, undertaken in the laboratory of Jacob Böeseken at the University of Delft in the Netherlands, involved adding various polyhydroxy compounds to aqueous boric acid solutions and studying the effect on the acidity and the conductivity of the resulting solu-

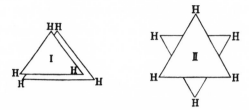

Fig. 12. Bischoff's projection formulas of ethane: I, eclipsed; II, staggered.

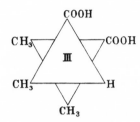

Fig. 13. Bischoff's projection formula of trimethylsuccinic acid.

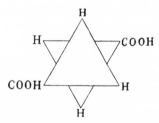

Fig. 14. Bischoff's projection formula of succinic acid.

tion. Van't Hoff had studied this effect qualitatively in 1908 and proposed that the increase in acidity when certain polyhydroxy compounds were added was due to the formation of a cyclic borate ester, which presumably was a stronger acid than boric acid. When Böeseken initiated these studies in 1913, he explained that glycerol had a greater effect than ethylene glycol in increasing the acidity of boric acid solutions (Fig. 15) because only in the former was it impossible for all the hydroxyl groups to exert their natural repulsion and thus prevent the coplanarity of the two reacting hydroxyl groups. The requirement of coplanarity for the reactive hydroxyl groups was confirmed by the

DIOL	Δ
HOCH₂CH₂OH	-1.0
CH₂CH(OH)CH₂OH	-0.7
HOCH₂CH(OH)CH₂OH	+11.9
cis-1,2- cyclopentanediol	+149.0
cis-1,2-cyclohexanediol	-6.2
trans-1,2-cyclohexanediol	-8.4
cis-1,2-cycloheptanediol	+137
trans-1,2-cycloheptanediol	+46

Fig. 15. Conductivity changes in aqueous boric acid solutions of some diols.

observation that only the cis isomer of 1,2-cyclopentanediol showed an increase in conductivity. Much to Böeseken's surprise, he then found that neither of the 1,2-cyclohexanediol isomers produced an increase in conductivity. Böeseken suggested that "the cyclohexane ring seemed to possess a certain flexibility which allowed the adjacent hydroxyl groups to obey to a certain extent their 'natural repulsion' and thus no longer remain in the same plane."[18]

One of Böeseken's students, H. G. Derx, continued these studies and found that both the *cis-* and the *trans*-1,2-cycloheptanediols exhibited an increase in conductivity (Fig. 15). Derx was able to provide a more satisfactory explanation for these observations by utilizing the ideas proposed by Sachse some thirty years earlier. Using special models that he had prepared for this purpose, Derx observed that when the cyclohexanediols were in the chair conformation, in neither the cis nor the trans diol were the two hydroxyls in the same plane. (*a* and *b* in Fig. 16 show the hydroxyl groups in axial-equatorial and diequatorial positions for the cis and trans isomers, respectively.)[19] Although the models clearly showed that it was possible to have the two hydroxyls in the same plane when the trans diol was in the boat conformation (*c* in Fig. 16), they apparently did not stay coplanar, because the flexibility of the boat permitted the hydroxyl groups to move away from each other (*d* in Fig. 16).

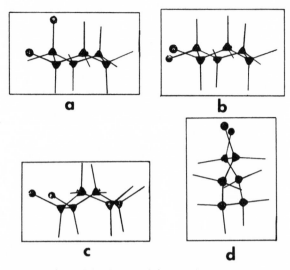

Fig. 16. Derx's models of 1,2-cyclohexanediols.

H. G. Derx, "Contributions à la connaissance de la configuration des sytèmes annulaires dans l'espace," *Recueil des Travaux Chimiques des Pays-Bas*, 41 (1922), 320. Reproduced by permission.

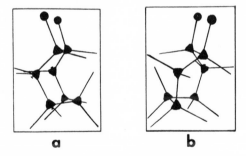

Fig. 17. Derx's models of 1,2-cycloheptanediols.

H. G. Derx, "Contributions à la connaissance de la configuration des systèmes annulaires dans l'espace," *Recueil des Travaux Chimiques des Pays-Bas*, 41 (1922), 328. Reproduced by permission.

An examination of models of *cis-* and *trans-*1,2-cyclohep-tanediol (*a* and *b* in Fig. 17) indicated that for both isomers there would be no difficulty in placing the two hydroxyl groups in the same plane. Why they should remain so if the 7-membered ring possessed any flexibility was not discussed.

Quantitative studies were continued by another student in

DIOL	$\underline{\Delta}$ [a]	K_{18} [b]
HOCH$_2$CH$_2$OH	-1.0	0.14
HOCH$_2$CH(OH)CH$_2$OH	+11.9	0.44
cis-1,2-cyclopentanediol	+149	10
cis-1,2-cyclohexanediol	-6.2	0.16
trans-1,2-cyclohexanediol	-8.4	-

[a] Conductivity changes.

[b] Equilibrium constants measured for diols and acetone.

Fig. 18. Data for some diols reacting with aqueous boric acid and with acetone.

Böeseken's laboratory, Peter Hermans, who undertook rate and equilibrium studies for the reaction of the diols with acetone.[20] He observed that even though compounds such as ethylene glycol and *trans*-1,2-cyclohexanediol had no measurable effect on the conductivity of boric acid solutions, they did react with acetone (Fig. 18). The study was then extended to the reaction of an acyclic compound, hydrobenzoin (1,2-diphenyl-1,2-ethanediol), with acetone. The rate and equilibrium constants, at two temperatures, for the forward and the reverse reactions were determined for the diastereomers (Fig. 19).

The differences in the rates of the diastereomers were explained as follows (with reference to Fig. 20): "The configurations with a minimum of potential energy will be those where the phenyl groups lie at the maximum distance from each other representing a projection in the direction of the central C-C bond. The energy required to bring the two OH groups closely together (into the 'favorable' position) will be much smaller in the case of the racemic isomer (*E*) than in that of the inactive isomer (*D*) because in the latter it is required to also bring the phenyl groups closely together."[21]

Hermans was able to show how the rate of the reaction was related to the fraction of the molecules in the favorable conformation. This kind of analysis did not appear again until thirty years later, in the publications of Saul Winstein and N.J. Holness, and Ernest Eliel and Carl Lukach.[22]

	$\underline{dl}$	$\underline{meso}$
$K_{eq}(25°)$	8.557	0.4402
$K_{eq}(49.8°)$	4.292	0.2662
Q(kcal/mole)	5.15	3.74
$10^3 k_1(25°)$, hr^{-1}	23.0	2.63
$10^3 k_1(45°)$, hr^{-1}	130.0	21.2
$10^3 k_2(25°)$, hr^{-1}	2.69	5.98
$10^3 k_2(45°)$, hr^{-1}	30.4	79.3
I_{for}(kcal/mole)	16.7	20.2
I_{rev}(kcal/mole)	21.9	23.9
ΔI	5.2	3.7

Q= Heat of Reaction   I = Activation Energy

Fig. 19. Summary of equilibria and rate measurements for reaction of 1,2-diphenyl-1,2-ethanediol with acetone.

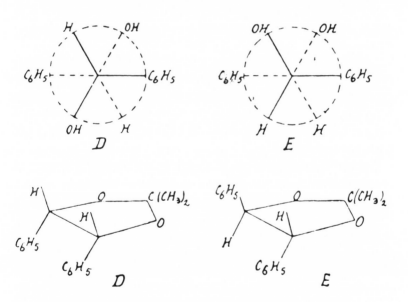

Fig. 20. Illustrations used by Hermans for diastereomeric diol reactants and ketal products.

P. H. Hermans, "Uber die Reaktion einiger Glykole mit Aceton," *Zeitschrift für Physikalische Chemie*, 113 (1924), 351. Reproduced by permission.

Hermans also became interested in quantitative calculations that could provide information about the relative stability of the cyclohexane conformers. Using the scale models devised by Derx and a mathematical approach similar to that utilized later by Barton, Hermans and a colleague, J. Berk, determined that the chair conformer was more stable than the boat. Because Hermans anticipated that no journal would accept an article that required such extensive typesetting for the mathematical formulas included in the manuscript, this work was never published except as part of Hermans' 1924 doctoral dissertation. Later Hermans made a brief reference to the calculations in an article concerning the conformational analysis of cycloheptane.[23] In this article he introduced conformational considerations similar to those encountered in the work of Prelog a decade later, and used the recently available Stuart space-filling models to illustrate the conformational features.

The importance of conformational analysis to the understanding of chemical and biochemical reactions was not lost on Hermans. "Perhaps some time (probably only in the distant future) a detailed knowledge of these subtle conformational [konfigurativen] differences will be of importance for the understanding of certain biochemical phenomena. In the finely tuned chemical behavior of living substances, the influence of conformational differences of the type considered might no doubt play an important role."[24]

The Nobel Laureate Prelog contributed to another dimension of the chemistry of many-membered ring compounds.[25] Prelog set about to demonstrate the relationship between ring size and conformation (he used the term *constellation*) and the magnitude of the equilibrium constants for a number of reactions. For example, to explain the equilibrium preference of cyclodecanone over its cyanohydrin addition product, Prelog suggested that the ketone has two conformations (illustrated with Stuart models)—one containing the carbonyl group inside, and the other outside, the ring. The one with the carbonyl group outside should form the cyanohydrin more easily. Unfortunately, in this instance Prelog proposed that any stability noted for the inside carbonyl was due to intramolecular hydrogen bonding from some of the internal ring hydrogens to the carbonyl oxygen.

The articles by Barton and Prelog signaled the birth of conformational analysis as a respected and valued field of study. Although the shift to dynamic stereochemistry was indeed dramatic, one can only wonder why it was so late in coming. It would seem that conformational analysis emerged suddenly because the time was ripe. Colin Russell, who summarized some of the obstacles to progress in this area, has concluded that the connecting links or keys to long-standing problems were "forged from an alloy of organic chemistry, physical chemistry and chemical physics." With this view Barton apparently concurred, as expressed in a letter to Hermans. "The question of why the work of this school [Böeseken's] was not well appreciated is an interesting one. I think for myself, that a theoretical treatment only becomes well accepted if there is a real need for a large body of chemists to use it. Thus my own paper in *Experientia* (1950) was in an obscure journal (for organic chemists) and yet it acted as a crystallization nucleus because steroid chemists had a great need of its generalizations. Cortisone chemistry was just beginning seriously in 1950 and there were thousands of chemists working on steroids."[26]

NOTES

1. James D. Watson, *The Double Helix* (New York, 1968), 38, 136.
2. D. H. R. Barton, "The Conformation of the Steroid Nucleus," *Experientia*, 6 (1950), 316, "Interactions Between Non-Bonded Atoms, and the Structure of cis-Decalin," *Journal of the Chemical Society*, (1948), 340–42, and "Molecular Models for Conformational Analysis," *Chemistry and Industry*, (1956), 1136–37.
3. O. Hassel, "The Cyclohexane Problem," *Tidsskrift for Kjemi, Begevesen og Metallurgi*, 3 (1943), 32 (Trans. in Norman L. Allinger and Ernest L. Eliot [eds.], *Topics in Stereochemistry* [New York, 1971], Vol. 6, p. 11); O. Bastiansen and O. Hassel, "Structure of the So-called cis Decalin," *Nature*, 157 (1946), 765.
4. Jacobus H. van't Hoff, *Voorstel tot nitbreiding der tegenword in de scheikunde grbruikte structur-formules in de ruimte* (Utrecht, 1874); J. A. Le Bel, "Sur les relations qui existent entre les formules atomiques des corps organiques et le pouvoir rotatoire de leurs dissolutions," *Bulletin de la Société Chimique de France*, 22 (1874), 337–47; Jacobus H. van't Hoff, *La Chimie dans l'Espace* (Rotterdam, 1875).

5. Jacobus H. van't Hoff, *Die Lagerung der Atome im Raume* (Braunschweig, Ger., 1877).

6. Adolf Baeyer, "Ueber Polyacetylenverbindungen," *Berichte der Deutschen Chemischen Gesellschaft*, 18 (1885), 2278.

7. August Kekulé, "Ueber die Constitution des Mesitylens," *Zeitschrift für Chemie*, 3 (1867), 214; William Henry Perkin, Jr., "First Pedler Lecture: The Early History of the Synthesis of Closed Carbon Chains," *Journal of the Chemical Society*, (1929), 1359.

8. Adolf Baeyer, "Ueber die Constitution des Benzols V," *Annalen der Chemie*, 258 (1890), 145–214, and "Ueber die Constitution des Benzols I," *ibid.*, 245 (1888), 178.

9. H. Sachse, "Ueber die geometrischen Isomerien der Hexamethylenderivate," *Berichte der Deutschen Chemischen Gesellschaft*, 23 (1890), 1365–66.

10. Ernst Mohr, "Die Baeyersche Spannungstheorie and die Struktur des Diamanten," *Journal für praktische Chemie*, 98 (1918), 315–53.

11. Walter Hückel, "Zur Stereochemie bicyclischer Ringsysteme," *Annalen der Chemie*, 441 (1925), 1–48.

12. Ossian Aschan, *Chemie der Alicyclischen Verbindungen* (Brunswick, Ger., 1905), 329.

13. Richard Moore Beesley, Christopher Kelk Ingold, and Jocelyn Field Thorpe, "The Formation and Stability of spiro-Compounds: Part I. *Spiro*-Compounds from *cyclo*-Hexane," *Journal of the Chemical Society*, 107 (1915), 1081; Jocelyn Field Thorpe, "Presidential Address," *ibid.*, (1931), 1022.

14. L. Ruzicka, "The Many-membered Carbon Rings," *Chemistry and Industry*, 13 (1935), 5.

15. Sterling B. Hendricks and Constant Bilicke, "The Space-Group and Molecular Symmetry of β-Benzene Hexabromide and Hexachloride," *Journal of the American Chemical Society*, 48 (1926), 3015; Roscoe G. Dickinson and Constant Bilicke, "The Crystal Structures of Beta Benzene Hexabromide and Hexachloride," *ibid.*, 50 (1928), 770.

16. van't Hoff, *La Chimie dans l'Espace*.

17. C. A. Bischoff, "Die dynamische Hypothese in ihrer Anwendung auf die Bernsteinsäuegruppe," *Berichte der Deutschen Chemischen Gesellschaft*, 24 (1891), 1085–95.

18. van't Hoff, *Die Lagerung der Atome im Raume*; J. Böeseken, "Über die Lagerung der Hydroxyl-Gruppen von Polyoxy-Verbindungen im Raum," *Berichte der Deutschen Chemischen Gesellschaft*, 46 (1913), 2612–28; J. Böeseken and J. van Griffen, "Sur les cyclohexanediols 1,2, et la souplesse de l'anneau benzenique," *Recueil des Travaux Chimiques des Pays-Bas*, 39 (1920), 186.

19. H. G. Derx, "Contributions à la connaissance de la configuration des systèmes annulaires dans l'espace," *Recueil des Travaux Chimiques des Pays-Bas*, 41 (1922), 312–42.

20. P. H. Hermans, "Über die Reaktion einiger Glykole mit Aceton," *Zeitschrift für Physikalische Chemie*, 113 (1924), 359.

21. Hermans, "Über die Reaktion einiger Glykole mit Aceton," 350–51.

22. S. Winstein and N. J. Holness, "Neighboring Carbon and Hydrogen: XIX. *t*-Butylcyclohexyl Derivatives. Quantitative Conformational Analysis," *Journal of the American Chemical Society*, 77 (1955), 5562–78; Ernest L. Eliel and

Carl A. Lukach, "Conformational Analysis: II. Esterification Rates of Cyclo-hexanols," *ibid.*, 79 (1957), 5986–92.

23. Barton, "Interactions Between Non-Bonded Atoms"; P. H. Hermans and C. J. Maan, "Beitrag zur Stereochemie des Siebengliedrigen Kohlenstoffringes," *Recueil des Travaux Chimiques des Pays-Bas*, 57 (1938), 643–52.

24. Hermans, "Über die Reaktion einiger Glykole mit Aceton."

25. V. Prelog, "Newer Developments of the Chemistry of Many-membered Ring Compounds," *Journal of the Chemical Society* (1950), 420–28.

26. Colin A. Russell, "The Origins of Conformational Analysis," in O. Bertrand Ramsay (ed.), *Van't Hoff–Le Bel Centennial* (Washington, D.C., 1975), 176; Derek H. R. Barton to Peter H. Hermans, Spring, 1972. Further details on early molecular models and on the development of stereochemical concepts can be found in O. Bertrand Ramsay, "Molecular Models in the Early Development of Stereochemistry," in O. Bertrand Ramsay (ed.), *Van't Hoff–Le Bel Centennial* (Washington, D.C., 1975), 74–96, and *Stereochemistry* (London, 1981).

A New Science and a New Profession: Sugar Chemistry in Louisiana, 1885–1895

BETWEEN 1885 and 1895 the Louisiana sugar industry experienced a scientific and technological revolution in methods, process apparatus, and scale of operations. The animal-powered mills and open kettles characteristic of the antebellum period were supplanted by large, technically designed, and scientifically controlled central factories. In 1880 there were approximately 1,000 sugar houses in Louisiana with an average annual production of 110 long tons of sugar per house. By 1900 fewer than 300 factories constituted the state's sugar industry, but yearly production averaged over 980 long tons for each sugar house. One commentator of the period, Mark Twain, described a modern Louisiana sugar factory as a "wilderness of tubs and tanks and vats and filters, pumps, pipes and machinery."[1]

This new industrial world that emerged in rural Louisiana was brought about in large part by a variety of local institutions working in alliance with the United States Department of Agriculture. They included the Louisiana Sugar Planters' Association, the Louisiana Sugar Experiment Station, the Audubon Sugar School, and Louisiana State University. These organizations facilitated the introduction of a progressive chemical and engineering technology, derived in part from the European beet sugar industry, into the traditional plantation culture of the Deep South.

The emergence of a scientific and technical institutional infrastructure was crucial to the development of the late nineteenth-century Louisiana sugar industry. Yet this complex transformation of manufacturing processes would never have taken place without the labor and innovations of a large number of experts working within the organizational structure. One group of experts, the sugar chemists, employed analytical skills and the knowledge of organic reactions to improve production efficiency in the sugar house.

To date, historians of American chemistry have generally focused their attention on the activities of either a few eminent scientists, like Theodore W. Richards, Gilbert N. Lewis, and Ira Remsen, or outstanding chemistry departments, like those at Johns Hopkins University, Harvard University, and the University of Chicago. Second- and third-level chemists and regional institutions have been almost totally neglected. Studies of the elite—the group of scientists at the very apex of the professional pyramid representing the American scientific community—are important indeed. Yet if we are truly to understand the progress of chemistry and chemical technology in America, we must also examine the lower tiers composing the bulk of investigative and commercial applications activity. For example, scholars must not neglect to examine the work of those creative chemists who made significant contributions to the development of the nineteenth-century agricultural and food industries. Their application of organic chemistry to the solution of practical problems was in part responsible for the drive to industrial maturity that occurred in the United States between 1875 and 1914.[2]

Between 1885 and 1895 the sugar chemist in Louisiana could be found plying his trade in either the experiment station, the university, or the sugar-house laboratory. His workplace and his professional world were far from tranquil. On the contrary, they were characterized by change and tension. In part, the challenges that confronted the chemist were the consequence of rapid technological changes. In order to maintain his employer's economically competitive position, the sugar chemist was constantly pressured to keep abreast of new developments in carbohydrate chemistry and process technology. Yet in addition to the everyday strains associated with his scientific and technical responsibilities, the sugar chemist's institutional setting and professional environment made his job even more trying. The chemist was a newcomer to the sugar-industry work force, and his ability to improve product quality and the efficiency of the manufacturing processes was often viewed with skepticism by the tradition-bound planter. Many Louisiana planters seriously questioned whether the application of scientific principles would ever lead to profits. In addition, the artisan sugar boilers, who previously had been charged with the responsibility of controlling the sugar-making process, correctly perceived the chemists as a threat to their livelihood and occupational status. On many occasions young university-trained chemists, unaccustomed to the coarse manners of the sugar-house floor, were shunned by the experienced and well-seasoned sugar boilers. And finally, many chemists experienced an inner struggle as well. Those sugar chemists who had extensive European educational training were forced to reconcile their vision of the German research ideal and their perceived role within the international scientific community with the reality that utility had priority over pure research and that regionalism, not internationalism, characterized their current working environment in Louisiana.

From the Age of Jackson to the 1880s, a few enlightened planters, along with a small group of medical men and consulting chemists, were the source of chemical expertise in the Louisiana sugar industry. By 1880, however, the Louisiana Sugar Planters' Association (LSPA), a group of the wealthiest and most politically powerful sugar planters in the state, recognized that the lack of scientific expertise could prevent it from successfully competing against the dynamic European beet sugar industry. To overcome this deficiency, this group pursued several strategies to secure university-trained scientists and engineers for their industry. As a result of the relationship it cultivated with the United States Department of Agriculture (USDA) and that existed between the two organizations between 1883 and 1889, the federal agency not only supplied the state with a large corps of formally trained research chemists but also facilitated the transfer of European developments in the chemical control of manufacturing processes and the introduction of new plant apparatus. The USDA's interest in the Louisiana sugar industry resulted in the use of systematic chemical analysis to monitor large-scale processes, as well as the introduction of new extraction, clarification, filtration, and evaporation apparatus.

Nevertheless, Louisiana planters regarded the USDA experiments as only moderately successful, since the agency's trials failed to establish conclusively the economic advantages of the new processes. The presence of the USDA in Louisiana and the unanswered questions from these trials convinced the LSPA leaders to sponsor a sugar experiment station. Through private subscription the LSPA established the Louisiana Sugar Experiment Station, which was first located in Kenner, Louisiana, in 1885.[3] William Carter Stubbs, from the Alabama Experiment Station, was hired as the LSPA station's first director.

Stubbs, who was director of the station between 1885 and 1905, proved to be popular and effective. Born near Williamsburg, Virginia, in 1843, he began his studies at the College of William and Mary in 1860. His studies were interrupted by a brief stint in the Confederate army, and after his capture and parole, he completed his preliminary studies at Randolph-Macon College and again served in the Confederate cavalry until the surrender at Appomattox. In 1865 Stubbs enrolled at the University of Virginia, where he completed the master's course in chemistry and geology in 1867. He then studied analytical chemistry under John William Mallet for one year. Mallet, who had been trained by Friedrich Wöhler at Göttingen, impressed upon Stubbs that chemistry was a quantitative rather than a qualitative discipline. Between 1869 and 1872 Stubbs served as professor of chemistry at Alabama Agricultural and Mechanical College at Auburn. He gained additional responsibilities in 1877, when he assumed the position of Alabama state chemist.[4]

Throughout most of his career, Stubbs's research focused upon field studies. Most significantly, his conception of agricultural chemistry was rooted in traditional American ideas and practices common between 1860 and 1880. His scientific activities in Louisiana were strongly influenced by Samuel W. Johnson's popular treatises, *How Crops Grow* and *How Crops Feed*.[5] Stubbs employed Johnson's views not only in formulating the research program at the Louisiana Sugar Experiment Station but also in teaching scientific agriculture to his students at Louisiana State University. Stubbs's understanding of plant growth, like that of Johnson, was based upon the analysis of inorganic ash residues and organic compounds, an approach that was the result of chemical and agricultural investigations first proposed by Justus Liebig in the early 1840s and employed extensively by European and American chemists during the late 1840s and early 1850s. Stubbs correlated plant growth stages with the relative quantities of inorganic acidic and basic salts found when various proportions of the plant were ashed. Furthermore, he estimated the volatile components of plants (nitrogen, carbon, hydrogen, oxygen, phosphorus, and sulfur) and extracted such loosely defined organic constituents as fibers, fats, and albuminoids.

During the late 1880s, the experiment station was moved from Kenner to Audubon Park in New Orleans. At about this time Stubbs began to reorient the experiment station's research program away from field trials and more toward sugar processing improvements. This change in emphasis was in part due to a shift in the planters' interest from the culture of sugarcane to the large-scale manufacture of raw sugar. In particular, the planters were debating the merits of the new diffusion process as a possible alternative to the crushing of cane during extraction. Having discovered that many of his assistants possessed a European education in chemistry, Stubbs sought to take advantage of their knowledge in carbohydrate chemistry to solve complex manufacturing problems. He designed manufacturing studies so that data from large-scale sugar production tests would also yield information about the formation of complex carbohydrates and invert sugars in the cane. The conversion of glucose to sucrose, and the complex interaction of gums, pectose sugars, and albuminoids, were chemical reactions that on one level were the basis of manufacturing problems, while on another level explained the biochemical development of cane from seed germination to maturity. Thus, fermentation, an ever-present threat to efficient processing, was closely tied to the

The "best looking fellow" at the Audubon sugar experiment station, New Orleans, 1894.

Courtesy of *The Charles Edward Coates Papers*, Special Collections, LSU Libraries, Louisiana State University and Agricultural and Mechanical College.

presence of albuminoids, and an understanding of these substances and their associated reactions partially revealed the nature of plant growth.

Stubbs's reliance upon the research efforts of the experiment-station chemists who had been trained in the fundamentals of organic structural chemistry led to the practice of a new type of agricultural chemistry at the Audubon experiment station during the 1890s. Investigators focused upon problems related to plant biochemistry, and they often employed new ideas and techniques originating in the European laboratories.

A number of chemists found employment at the station between 1888 and 1905, including some, like T. H. Jones, Maurice Bird, and W. Wipprecht, who had been trained at southern universities. Wipprecht, however, had also been educated at Göttingen, and he proved to be the first of a large number of Bernard Tollens' former students who applied their theoretical knowledge of sugar chemistry to the solution of practical problems at the Louisiana Sugar Experiment Station.[6]

Tollens, who had received his training under Rudolf Fittig at the University of Göttingen, Emil Erlenmeyer at the University of Heidelberg, and Charles Adolphe Wurtz at the University of Paris, had succeeded Wilhelm Wicke in 1873 as professor and director of the laboratory at the Agricultural Institute of Göttingen. Historians of chemistry have neglected Tollens' career in science, perhaps because he was the practical chemist working out in the laboratory the structural theories proposed by better known and more illustrious contemporaries. Nevertheless, Tollens had a central position within the nineteenth-century German community of chemists, frequently interacting with industrial concerns and academic scientists over scientific problems of mutual interest. Until 1911 Tollens and his students focused their efforts on determining the exact chemical nature of the unknown carbohydrates contained in the crude fibers and nitrogen-free extracts of crops. Emil Fischer's work on the structure of carbohydrates enabled Tollens to understand basic carbohydrate reactions like hydrolysis, hydrazone formation, and fermentation.[7] With this knowledge Tollens successfully separated and analyzed many carbohydrate compounds normally found in plant materials, including hexosans, pentosans, and methyl pentosans. He also conducted studies on enzyme reactions and the process of fermentation. Similar studies became the focus of the Louisiana Sugar Experiment Station after 1890, when Tollens' former students Charles A. Browne, Fritz Zerban, Peter A. Yoder, and William E. Cross applied their knowledge of carbohydrate chemistry and structural chemistry to the solution of practical sugar manufacturing problems.

Although a majority of the chemists with Ph.D. degrees working at the station between 1889 and 1910 were former students of Tollens, graduates of Harvard, Johns Hopkins, and Zürich Polytechnical Institute also made valuable research contributions. Of the two station chemists educated at Harvard—Horace Everett Lincoln Horton and Josiah Thomas Crawley—Crawley studied agricultural and physiological chemistry under Walter Maxwell, who in 1893 replaced his student at the Audubon Sugar School.[8] Also in 1893 Stubbs's staff was enriched by the addition of Jasper L. Beeson, a chemist trained by Remsen. Although manufacturing problems and questions on plant growth remained central, Beeson and Maxwell used their talents, with Stubbs's approval, to redirect the nature of station investigations.

Beeson was born in Alabama in 1867. After graduating from the University of Alabama in 1889, he was appointed instructor of physics at the university, and a year later he found employment as a chemist for the Alabama Geological Survey. In 1891 he entered Johns Hopkins, where he earned a Ph.D. degree in 1893.[9] After joining the staff at the Louisiana Sugar Experiment Station, he conducted research and taught chemistry to students of the station's sugar school.

In his research Beeson investigated the link between problems in sugar manufacture and the processes of plant growth and nutrition. The most controversial topic among Louisiana planters and sugar manufacturers during the early 1890s was whether diffusion or milling was more economical. Since it had been noticed that juices obtained from two successive milling operations were completely different in color, it was postulated that the material contained in the cane internodes was different from that expressed from the nodes during the second milling. Beeson found that the substances extracted from the nodes were highly colored, giving a heavy precipitate when mixed with a solution of subacetate of lead and coagulating when heated to the boiling point. However, extracts from the internodes were clear and light in color, produced little coagulation upon heating, and contained more albuminoids and reducing sugars than the nodes. In response to these findings, Beeson hypothesized that the physiological function of cane nodes was similar to that of seeds in flowering plants—storing food for the sustenance of the young plant during that initial period before it had taken root sufficiently to draw nutrients from the earth. In Beeson's experiments on pedigreeing cane, he chemically analyzed sugarcane grown from seed tops, middles, and butts and found no significant differences. However, he discovered the presence of less sugar and more solids—albuminoids, amide nitrogen, and ash—in the tops. Therefore, he suggested using tops for seeds—cutting down to the first joint that had cast its leaves and sending the remaining sugar-rich portion to the mills for processing.[10]

Maxwell also studied fundamental problems in plant biochemistry that had their origin in manufacturing. Born in Great Britain in 1854, Maxwell studied at the City and Guilds Institute at South Kensington. Later he worked with Ernst Schulze at Zürich Polytechnical Institute, where he investigated the chemistry of plant cell membranes and the constituents of leguminous seeds.[11] In 1888 Maxwell enrolled at Harvard, where he conducted a course in physiological chemistry. Between 1889 and 1893 he served the USDA as the assistant chemist supervising the agency's beet sugar experiments at Schuyler, Nebraska. After accepting a position at the Audubon Park sugar station in 1893, Maxwell remained in Louisiana until his appointment in 1895 to the directorship of the Hawaii Sugar Experiment Station.

Maxwell's investigations in Louisiana were conducted in response to problems encountered during diffusion-process trials at the station. He hypothesized that the presence of noncrystallizable organic bodies often interfered with the crystallization of diffusion juices. Diffusion, he thought, also extracted substances that were normally left in the bagasse during milling and left behind other substances usually extracted by pressure treatment.[12] In the past, nitrogen analyses conducted at the station had been based on the assumption that all nitrogen was in the form of albuminoids. However, Maxwell performed a Kjeldahl nitrogen assay on mill and diffusion juices to determine total nitrogen, and then he estimated amide nitrogen by the Stutzer method. Maxwell determined that whereas the nitrogen in mill juices was distributed between albuminoids and nonalbuminoids (chiefly in the form of amides) in the ratio of one to two, that in diffusion juices it was distributed in the ratio of one to three. Since albuminoid bodies were necessary for the formation of the coagulation blanket during the clarification process, it was now evident why hot water diffusion was often accompanied by crystallization and clarification problems.

Like his colleague Beeson, Maxwell related his findings to physiological interpretations of cane growth. He claimed that, in mature seed, nitrogen was primarily in an albuminoid form; after the seed germinated, nitrogen was converted to the water-soluble amide form. Thus, amide compounds, along with glucose, were easily transported to the various regions of the plant, where they furnished the raw materials for growth while being reduced to albuminoid form once again.

Maxwell's investigations opened new areas in sugar chemistry. Yet his work had practical consequences as well. His analytical studies of nitrogen compounds effectively marked the end of diffusion trials in Louisiana. Although the diffusion process had proven itself an effective method of obtaining high extraction yields, the addition of excess water resulted in increased fuel consumption. This dilution problem could have been surmounted, perhaps, by the use of efficient multiple-effect evaporation apparatus. However, chemical process problems, particularly during crystallization, proved to be the process's death knell.[13]

Thus, between 1890 and 1895 the research staff at the Louisiana Sugar Experiment Station concentrated on problems in sugar chemistry that were related not only to manufacturing difficulties but also to important questions in plant physiology and nutrition. This new type of agricultural chemistry practiced at the experiment station at Audubon Park was quite different from the inorganic analyses that Stubbs had conducted at the Kenner location. Unlike Stubbs, Beeson and Maxwell were apparently not satisfied with reporting the results of their work only in the experiment station's bulletins and the *Louisiana Planter and Sugar Manufacturer*. These chemists were eager not only to prove their work to the skeptical planters but also to gain the recognition of the scientific community. Therefore, it is not surprising that Beeson communicated his findings in the *Journal of the American Chemical Society* nor that Maxwell published his results in the *Sugar Cane*.[14]

Between 1885 and 1895, both the station chemists and the factory chemists wanted to demonstrate their legitimacy within the sugar industry. In particular, the factory chemists wanted to overcome both the skepticism of their employers and the criticisms of the artisans whom they were gradually displacing from the sugar houses. The sugar chemists in Louisiana viewed organization as the best strategy for securing their professional goals.

Factory chemists led an early movement to organize a professional association for sugar chemists in Louisiana. In 1889 Lezin Becnel, a chemist employed at the McCall Bros. Plantation in Ascension Parish, took the initiative to establish a local chemists' society. He acted primarily in response to the perceived widespread opposition to sugar chemists by both skilled and unskilled laborers. Becnel hoped that an association would clearly define the role of the chemist in the sugar industry as well as assist in uniting chemists with their planter employers for their mutual benefit. Becnel asserted that the factory chemist was a misunderstood, frequently maligned professional whom plant workers often called a "crank" working in a "drugstore."[15]

While factory chemists were promoting and clarifying their position, station chemists were recognizing that the scientists whose work was guided by theory suffered from a lack of prestige within the planter circles. Louisiana planters emphasized that their business needed the contributions of *practical* chemists. According to this view, the successful chemist possessed "a large stock of practical knowledge, and of experience in actual sugar house work. . . . To employ a professor of chemistry unpossessed of this experience would be to lose a season."[16] Thus, both industrial and station chemists—two groups of chemists possessing somewhat different educational backgrounds and professional objectives—had sufficient reasons to unite.

On June 15, 1889, the Louisiana Sugar Chemists' Association

held its first meeting at the Audubon sugar experiment station.[17] Some members of the station's staff, including Stubbs, attended, and Bennett Battle Ross from the Baton Rouge station was elected the group's first president. The association's objectives included the dissemination of chemical knowledge, the discussion and evaluation of analytical methods, and the adoption of a standard system of factory chemical-control statistics. Membership in this organization was limited to chemists who were either employed in the analysis of agricultural products or engaged in scientific agriculture. The organization was led by its officers and two permanent committees of three members each. The committee on statistics was selected from practical sugarhouse chemists, whereas the committee on analysis was composed of station chemists.

During the summer of 1889, led by Ross and Bird of the sugar station, the committee on analysis compiled a list of the most satisfactory analytical procedures for the determination of sugar. The fact that they published their data in a widely circulated pamphlet that fall indicated that these station chemists were attempting to establish themselves as experts within the Louisiana community of chemists. The *Report of the Committee on Methods of Sugar Analyses of the Louisiana Sugar Chemists' Association* outlined in a simple manner the methods and apparatus necessary for the determination of sucrose, specific gravity, density, total solids, glucose, ash, and fiber. This detailed description was written for the benefit of analysts who were "just entering . . . or [had] only a limited experience in the laboratory of the sugar house."[18]

Thus the committee established itself as an authority in approving the numerous chemical procedures used in the chemical control of the factory. In addition, the committee of station chemists conducted precise polariscopic studies to ascertain the absolute value of the constant employed in Clerget's method for the determination of sucrose in the presence of glucose.[19] This investigation obviously had practical significance to the Louisiana sugar industry, which processed millions of pounds of sugar annually. Even a small variation in Clerget's constant could

affect the profits of large-scale processors. However, the extreme
care taken with reagents and experimental conditions, as well as
the treatment of data, suggests that Ross and Bird were also
hoping to earn the respect of the scientific community. They conducted their experiments utilizing a double-compensating po-
lariscope in a 31°F room at the New Orleans Cold Storage Company. The results were compared with data obtained from gravi-
metric determinations using Fehling's solution. Indeed, the
committee was pursuing work begun by Hans Heinrich Landholt in Germany, and it published its results in the *Journal of
Analytical Chemistry.*

Concurrently, the association's statistics committee began to
devise a standard form for reporting chemical-process information. It was hoped that these sheets would eventually become
the basis for comparing the performance of various Louisiana
sugar houses and for indicating manufacturing inefficiencies.
Becnel asserted that in the future "our records should also include the cost of every item, from the cutting of the cane to the
cost of laying down one pound of sugar on our plantation land-
ings, so as to . . . reduce the actual cost of manufacture."[20] The
introduction of chemical control into the Louisiana sugar industry ultimately revolutionized business practices. For example,
after St. Mary's Parish planter John N. Pharr employed a chemist at his Glenwild factory, annual reports became more detailed
and manufacturing costs were included. The chemists' division
of the manufacturing process into units later became the basis
for cost accounting.

The Louisiana Sugar Chemists' Association met in the fall
and the spring, and discussions usually consisted of works-in-
progress reports by various members. Concurrent with the
movement to stimulate research within the membership, the
chemists' association began to take steps toward identifying itself within a larger scientific community. In 1891 the organization expressed interest in joining a national group of chemists
led by Albert C. Peale. By 1893 most of the members had already
joined the American Chemical Society (ACS), and in late 1893
the association officially took steps to join that body. In 1894 the

Louisiana Sugar Chemists' Association became a part of the newly formed New Orleans section of the ACS. This local ACS branch was short-lived, however, operating only until 1898.[21]

The efforts of chemists to advance the utility of their profession gained widespread acceptance in Louisiana. The factory chemist became an integral part of sugar manufacturing during the 1890s. To remain competitive, planters were forced to rely upon chemical-control data to minimize process losses and to maximize product quality. However, the status of a research-oriented station chemist never became as secure as that of his factory counterpart. With the repeal of the Bounty Law in 1894, which had levied stiff tariffs on imported sugars, the American market was flooded with cheap imported sugar, and the Louisiana sugar industry began a gradual decline. This economic situation was not conducive to the support of carbohydrate research like that conducted by Beeson and Maxwell. To a large degree, planters saw the answer to their problems not in the study of complex organic substances but in political activities. The redirection of the planters' interest resulted in the decline of first-rate research at the Louisiana Sugar Experiment Station, as pure science assumed a much lower priority than theretofore.

Thus, the world of the sugar chemist in nineteenth-century Louisiana was characterized by change and tension. The agricultural chemistry of Liebig, practiced by Stubbs at the Louisiana Sugar Experiment Station, was unable to solve manufacturing problems that were arising from a new sugar technology. A new kind of American chemist, like Maxwell and Beeson, applied modern organic chemistry to the solution of complex processing problems. Yet these European-trained scientists had to deal with both skepticism and inner conflict. Along with the factory chemists, the station chemists saw organization as a way of overcoming their tensions. Although the utility of chemical control was gradually accepted, a change in economic conditions adversely affected the aspirations of the research-minded station chemists in Louisiana. After 1895 the experiment-station chemists could hope to pursue research only if funded by the federal government. The story of the gradual decline in the quality of research at the Louisiana Sugar Experiment Station is to be told on another occasion.

NOTES

1. Samuel L. Clemens [Mark Twain], *Life on the Mississippi* (New York, 1917), 384.
2. Harold Vatter, *The Drive to Industrial Maturity* (Westport, Conn., 1975).
3. William Carter Stubbs, "An Experimental Farm for the Promotion of the Sugar Interest" (MS in William Carter Stubbs Papers, Southern Historical Collection, University of North Carolina at Chapel Hill), and "Charter of the Louisiana Scientific Agricultural Association" (Printed copy in *ibid.*).
4. Stubbs's education was traced in a letter from William Carter Stubbs to Charles Albert Browne, February 27, 1922, in the Charles Albert Browne Papers, Library of Congress. Stubbs's scientific and technical contributions can be best reviewed by examining his many addresses and papers appearing in *Louisiana Planter and Sugar Manufacturer.* His monographs include *Sugar Cane. A Treatise on the History, Botany and Agriculture of Sugar Cane, and the Chemistry and Manufacture of its Juices into Sugar and Other Products* (New Orleans, 1897), *Sugar Cane* (Boston, 1903), and with Herbert Myrick, *The American Sugar Industry: A Practical Manual on the Production of Sugar Beets and Sugar Cane, and on the Manufacture of Sugar Therefrom* (New York, 1899). See also William Carter Stubbs, "Notes on Chemical Analysis" (MS in William Carter Stubbs Papers, College of William and Mary, Williamsburg).
5. Samuel W. Johnson, *How Crops Grow. A Treatise on the Chemical Composition, Structure and Life of the Plant, for All Students of Agriculture* (New York, 1887), and *How Crops Feed. A Treatise on the Atmosphere and the Soil as Related to the Nutrition of Agricultural Plants, with Illustrations* (New York, 1882). For information on Johnson, see Margaret Rossiter, *The Emergence of Agricultural Science: Justus Liebig and the Americans* (New Haven, 1975), 127–48.
6. For information on Jones, see *Louisiana Planter and Sugar Manufacturer,* 1 (1888), 107; C. A. Browne, "Bernard Tollens (1841–1918) and Some American Students of Agricultural Chemistry," *Journal of Chemical Education,* 19 (1942), 253–59. Tollens' work is briefly described in Aaron J. Ihde, *The Development of Modern Chemistry* (New York, 1964), 344–45.
7. To understand Tollens' research, see Bernard Tollens, *Kurzes Handbuch der Kohlenhydrate* (2 vols.; Breslau, 1895); W. E. Cross and B. Tollens, "Versuche über das Verhalten der Pentosen in garenden Mischungen," *Journal für Landwirtschaft,* 59 (1911), 419–28.
8. H. E. L. Horton, "Some Notes on the Determination of Sugars with Fehlings Solution," *Journal of Analytical Chemistry,* 4 (1890), 370–81; J. T. Crawley, "A Simplified Fat-Extracting Apparatus," *American Chemical Journal,* 11 (1889), 507–508.
9. Jasper Luther Beeson, "A Study of the Action of Certain Diazo-Compounds on Methyl and Ethyl Alcohols Under Varying Conditions" (Ph.D. dissertation, Johns Hopkins University, 1893).
10. J. L. Beeson, "A Study of the Constituents of Nodes and Internodes," "The Es-

timation of Crude Fibre in Sugar Cane," "Pedigreeing of Cane—'Tops from Tops,'" and "Effects of Fertilizers Upon Sugar Cane," in *The Chemistry of Sugar Cane and Its Products*, Louisiana Sugar Experiment Station Bulletin, 38 (Baton Rouge, 1895), 1341–71.

11. C. A. Browne, "Dr. Walter Maxwell," *Facts About Sugar*, 27 (1932), 24. One can readily ascertain the influence of Schulze upon Maxwell by examining the latter's early publications. See W. Maxwell, "On the Solubility of the Constituents of Seeds in Prepared Solutions of Ptyalin, Pepsin, and Trypsin," *American Chemical Journal*, 11 (1889), 354–57, "On the Presence of Sugar-Yielding Insoluble Carbohydrates in Seeds," *ibid.*, 12 (1890), 51–60, "On the Soluble Carbohydrates Present in the Seeds of Legumes," *ibid.*, 12 (1890), 265–69, "On the Methods of Estimation of Fatty Bodies in Vegetable Organisms," *ibid.*, 13 (1891), 13–16, and "On the Behavior of the Fatty Bodies, and the Role of the Lecithines, During Normal Germination," *ibid.*, 13 (1891), 16–24. Maxwell's interest in nitrogen compounds is reflected in "On the Nitrogenous Bases Present in the Cotton Seed," *ibid.*, 13 (1891), 469–71. His early studies on germination were published in "Movement of the Element Phosphorus in the Mineral, Vegetable, and Animal Kingdoms, and the Biological Function of the Lecithines," *ibid.*, 15 (1893), 185–95. Maxwell's USDA publications include: Walter Maxwell and Harvey W. Wiley, *Experiments with Sugar Beets in 1892*, USDA Division of Chemistry Bulletin, 36 (Washington, D.C., 1893) and *Experiments with Sugar Beets in 1893*, USDA Division of Chemistry Bulletin, 39 (Washington, D.C., 1894).

12. Walter Maxwell, "Organic Solids Not Sugar in Cane Juices," "Sulphurous Acid, Acid Phosphate and Lime as Clarifying Agents," and "Fermentation of Cane Juices," in *The Chemistry of Sugar Cane and Its Products*, Louisiana Sugar Experiment Station Bulletin, 38 (Baton Rouge, 1895), 1371–1408.

13. "Milling v. Diffusion," *Sugar Cane*, 27 (1895), 428.

14. See J. L. Beeson, "Notes on the Estimation of Crude Fiber in Sugar Cane," *Journal of the American Chemical Society*, 16 (1894), 308–13, "A Simple and Convenient Extraction Apparatus for Food-Stuff Analysis," *ibid.*, 18 (1896), 744–45, and "A Study of the Clarification of Sugar Cane Juice," *ibid.*, 19 (1897), 56–61.

15. "Louisiana Chemists Organizing," *Louisiana Planter and Sugar Manufacturer*, 2 (1889), 270. To understand Becnel's work, see Lezin A. Becnel, *Report on the Results of Belle Alliance, Evan Hall and Souvenir Sugar Houses, for the Crop of 1888* (New Orleans, 1889), and "General Plantation and Sugar House Statistics, the Manner of Keeping Same, and the Necessary Chemical Control," *Louisiana Planter and Sugar Manufacturer*, 2 (1889), 286–87.

16. "Notes and Comments," *Louisiana Planter and Sugar Manufacturer*, 3 (1889), 370.

17. "The Sugar Chemists," *ibid.*, 2 (1889), 283.

18. *Report of the Committee on Methods of Sugar Analysis of the Louisiana Sugar Chemists' Association* (Baton Rouge, n.d. [1889]), 201, reprinted in *Journal of Analytical Chemistry*, 4 (1890), 1–19.

19. The proposed revision of the constant used in Clerget's method caused a controversy among analytical chemists. See "Changes of Methods of Analyses by the Association of Official Agricultural Chemists," *Journal of Analytical*

Chemistry, 6 (1892), 259–62, and "Concerning the Constant to Be Used in Clerget's Inversion Process," *ibid.*, 6 (1892), 432–35, 519–24, 633–36. Clerget's procedure, first proposed in 1849, enables the analyst to determine the percentage of sucrose in a solution containing other optically active substances. Cane sugar (sucrose) is hydrolyzed by either acid or the enzyme invertase to a 1:1 mixture of glucose and fructose, which rotates polarized light in the opposite direction. The amount of change in rotation is proportional to the concentration of sucrose. For additional information on Clerget's method, see C. A. Browne and F. W. Zerban, *Physical and Chemical Methods of Sugar Analysis* (New York, 1941), 402–405.

20. Becnel, "General Plantation and Sugar House Statistics," 287.
21. Between 1896 and 1898 the inside cover of the *Journal of the American Chemical Society* listed a local New Orleans chapter. In 1900 the New Orleans section was no longer included, and in 1906 a second New Orleans chemists' association affiliated with the ACS was established.

The Development of a Scientific Community: Physical Organic Chemistry in the United States, 1925–1950

ASIDE from the major elements of structural theory, the theory of organic chemistry developed primarily during the twentieth century. Physical organic chemistry—literally, the application of the theory and methods of physical chemistry to organic chemistry but, in reality, the development of the theory of organic chemistry—grew slowly over the first three decades of this century, and particularly slowly in the United States. However, with the elucidation of the electronic structure of the atom, the development of a new theory of bonding, and the development of mathematical methods for describing atomic structure and bonding, the belief that organic reactions could be understood at a molecular level and that reasonable correlations could be made between structure and reactivity emerged.[1]

A number of early workers, like Moses Gomberg, Julius Stieglitz, Solomon F. Acree, Arthur Michael, and John Nef, could be considered the first generation of physical organic chemists in the United States. However, before 1930 there was certainly no community of chemists interested in organic theory and no hint that one would soon emerge. The early development of physical organic chemistry in the United States took place primarily at four institutions: Harvard University, Columbia University, the University of Chicago, and the California Institute of Technology–University of California, Los Angeles. Other strong programs in organic chemistry existed—at Johns Hopkins University (before 1941), the University of Illinois, the University of Wisconsin, Iowa State College, Duke University—but none had the same impact on the development of physical organic chemistry as those four. Before 1932 there was no exchange of personnel among the programs in the four institutions, though Cal Tech made more than one attempt to hire James Bryant Conant away from Harvard, and there was probably little, if any, informal exchange of information. Essentially, these programs in physical organic chemistry developed independently of one another.

What was unique about Harvard, Columbia, the University of Chicago, and Cal Tech–UCLA? To aid its young faculty members in their research efforts, each of these institutions provided a particularly favorable environment by way of encouragement, some freedom, funds and equipment, and students or research assistants. In addition, there were young chemists about who were willing to challenge the authority of classical organic chemistry. And most important, among the organic chemists and physical chemists there existed a willingness—indeed, a feeling that it was necessary—to exchange ideas.

Figure 1 lists the chemists who were the forefathers of the physical organic chemistry programs at Harvard, Columbia, the University of Chicago, and Cal Tech. Elmer Peter Kohler, who was from the Pennsylvania Dutch country, studied at Muhlenberg College and eventually took a Ph.D. degree with Ira Remsen at Johns Hopkins. After a fairly distinguished career as professor at Bryn Mawr College (Louis Fieser and Arthur Cope were among others who subsequently held the Bryn Mawr position), he moved on to Harvard in 1912 at the age of forty-five, replacing Charles Loring Jackson, who had been forced into retirement by illness. The dominant figure in the Harvard chemistry department was the physical chemist Theodore W. Richards. Although Richards' best work was behind him, the Harvard faculty consisted primarily of physical chemists, many of them Richards' own students. Kohler, then, could not help but discuss chemistry with physical chemists. He could, however, also discuss chemistry with Arthur Michael, who was on the Harvard staff from 1912 to 1936. Michael's role at Harvard is unclear, but there is little question that he was one of the early theorists in organic chemistry and that he was a strong personality who had some influence wherever he was.[2]

Kohler's chemistry was good, but it was his teaching that was outstanding. He had a profound influence on everyone who passed through his course in advanced organic chemistry. He had a strong interest in the detailed mechanisms of organic reactions, and many of his students apparently inherited his interest and enthusiasm. Perhaps his finest collaborative effort with a physical chemist was the training of Conant, who took his Ph.D. degree under the joint sponsorship of Kohler and Richards. Besides Conant, other students who passed through Kohler's course or his research laboratory included Frank Whitmore, Louis Hammett, Paul Bartlett, Arthur Cope, Max Tishler, Frank Westheimer, Stanley Tarbell, and Charles Price.

John "Pop" Nelson, a midwesterner, took his Ph.D. degree at Columbia with Marston Bogert, an old-line "classical" organic chemist. He did not, however, follow in the footsteps of his mentor. Nelson's chemical interests were eclectic. From 1910 to 1920, he wrote several influential articles on chemical bonding, most of them with K. George Falk, who had worked in Arthur A. Noyes's Research Laboratory of Physical Chemistry at the Massachusetts Institute of Technology. Another collaborator in these efforts was Hal Beans, a physical chemist at Columbia. Nelson was the first to use the hydrogen electrode in a biochemical problem, and later in his career he worked on the isolation of pure enzymes and the mechanism of enzyme action. Nelson, like Kohler, taught the advanced course in organic chemistry and had a strong interest in the mechanisms of organic reactions. Also like Kohler, Nelson had a profound influence on Louis Hammett.[3]

At the University of Chicago, the two forefathers of physical organic chemistry were John Ulric Nef and Julius Stieglitz. Nef was born in Europe but grew up in Massachusetts. He took his undergraduate degree at Harvard and then went to Germany, where he took his Ph.D. degree with Adolf Baeyer. After a short stay at Purdue University and then at Clark University, he was lured to the new University of Chicago in 1892 to head the chemistry department. He is best known for the Nef reaction, but he spent much of his career working on bivalent carbon and on the base-catalyzed isomerizations and oxidations of sugars. Among

Before 1920

HARVARD

E. P. Kohler
(1912–1938)

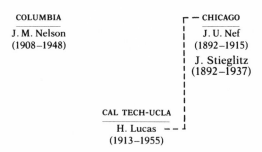

COLUMBIA

J. M. Nelson
(1908–1948)

CHICAGO

J. U. Nef
(1892–1915)

J. Stieglitz
(1892–1937)

CAL TECH-UCLA

H. Lucas
(1913–1955)

Fig. 1. Key for this and all subsequent figures: The dates record the period of work by the individual at the institution. A solid line connects an individual with his doctoral mentor or doctoral institution. A broken line connects an individual with an institution that he attended without receipt of a doctoral degree or with his postdoctoral mentor, or it implies some other loose connection described in the text.

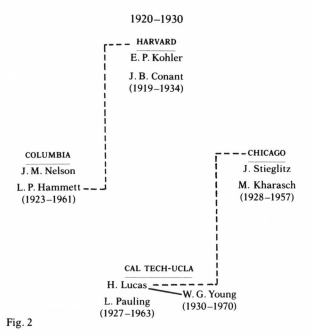

1920–1930

HARVARD

E. P. Kohler

J. B. Conant
(1919–1934)

COLUMBIA

J. M. Nelson

L. P. Hammett
(1923–1961)

CHICAGO

J. Stieglitz

M. Kharasch
(1928–1957)

CAL TECH-UCLA

H. Lucas
W. G. Young
L. Pauling   (1930–1970)
(1927–1963)

Fig. 2

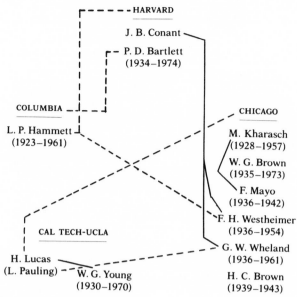

Fig. 3

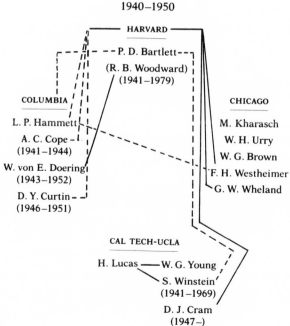

Fig. 4

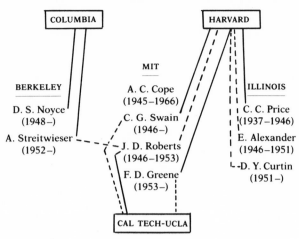

1945–1955

COLUMBIA   HARVARD

                    MIT

BERKELEY   A. C. Cope   ILLINOIS
                    (1945–1966)
D. S. Noyce   C. C. Price
(1948–)   C. G. Swain   (1937–1946)
                    (1946–)
A. Streitwieser   J. D. Roberts   E. Alexander
(1952–)   (1946–1953)   (1946–1951)

                    F. D. Greene   -D. Y. Curtin
                    (1953–)   (1951–)

                    CAL TECH-UCLA

Fig. 5

Elmer Peter Kohler

Biographical Memoirs, 27 (1952), facing p. 265. Reproduced by permission.

John Ulric Nef

Photograph by Cox of Chicago, from *Biographical Memoirs*, 34 (1960), facing p. 204. Reproduced by permission.

Julius Stieglitz

Journal of the American Chemical Society Obituaries, 60 (1938), facing p. 3.

Howard Johnson Lucas

Photograph by Ballam-Wanek-King Studio, from *Biographical Memoirs*, 43 (1973), facing p. 163. Reproduced by permission.

James Bryant Conant

Photograph by Josef Karsh, from *Biographical Memoirs*, 54 (1983), p. 90.

Paul D. Bartlett

William A. Pryor (ed.), *Organic Free Radicals*, American Chemical Society Symposium Series 69 (Washington, D.C., 1978), xiv. Reproduced by permission.

Morris Selig Kharasch

Photograph by the Llewellyn Studio, from *Biographical Memoirs*, 34 (1960), facing p. 123. Reproduced by permission.

Saul Winstein

Biographical Memoirs, 43 (1973), facing p. 321. Reproduced by permission.

his students who had a direct or an indirect effect on physical organic chemistry were Acree, who was at Johns Hopkins for a while; William Evans, who spent his career at Ohio State University; and Herman Spoehr, who was at the Carnegie Institute in Carmel, California.[4] Howard Lucas took a master's degree with Evans, and William G. Young worked with Spoehr before taking his Ph.D. degree. The California school of physical organic chemistry descends directly from Nef and the University of Chicago.

Although he was born in New Jersey, Stieglitz received his high school and university training in Germany. After taking his Ph.D. degree with Ferdinand Tiemann, he worked with Nef for a few months at Clark University and then went to work for Parke-Davis in Detroit. Whereas Nef went to the University of Chicago as head of the chemistry department in 1892, Stieglitz went as an unpaid docent. However, by 1907 Stieglitz was a full professor, offering courses called "Physical Chemistry Applied to Organic Problems" and "Physico-organic Research."[5] Stieglitz had a strong bent for physical chemistry, and he was apparently close to the two physical chemists in the department, Herbert McCoy and Alexander Smith. His research program included work on the Beckmann and Hofmann rearrangements, the catalysis of saponification reactions, stereoisomerism, color theory, the electronic theory of bonding, and, late in his career, pharmaceuticals.

When Nef died in 1915, Stieglitz became chairman of the department (he had unofficially been running it for several years), a post he held until his retirement in 1933. After his retirement, he continued to teach chemistry courses and may have had some influence on the development of the department in the 1930s.

Of the forefathers shown in Figure 1, only Stieglitz and Nef took doctorates in Germany. One might have expected chemistry at the University of Chicago to fit into the classic German mold, but of the four institutions shown, it probably has the longest tradition in physical organic chemistry. By the late 1930s, the Chicago department had a higher concentration of physical organic chemists on its staff than any other university department in the world.

In view of Gilbert Newton Lewis' contributions to the development of physical organic chemistry—the electronic theory of bonding and his acid-base theory, to name but two—it is only natural to look to the University of California at Berkeley for the beginnings of physical organic chemistry on the West Coast. However, from 1913 to 1937 no outsiders were hired to the chemistry staff at Berkeley. Therefore, the "organic chemists" in the department before the middle 1940s—Gerald E. K. Branch, Charles Walter Porter, and T. D. Stewart—were trained as physical chemists. Except for Melvin Calvin, they made no major contributions to organic chemistry or physical organic chemistry, and they certainly trained no students of note in either area. Even the book by Branch and Calvin, *The Theory of Organic Chemistry*, came at a time when physical organic chemistry had already become a thriving specialty in the United States.[6] In a sense, Lewis, a physical chemist, so dominated the intellectual climate at Berkeley that "things organic could not grow."

In 1913 the Throop College of Technology in Pasadena hired Lucas as an instructor.[7] After Lucas had taken his master's degree at Ohio State with Evans, he went to the University of Chicago to work for either Nef or Stieglitz, but the death of his father forced him to leave in 1910 without a Ph.D. degree. He was working as a chemist for the United States Department of Agriculture when he was hired by Throop.

Throop was officially renamed the California Institute of Technology around 1920. Noyes was brought from MIT to build a strong chemistry department, which he proceeded to do, building it much in the mold of his Research Laboratory of Physical Chemistry at MIT, with the faculty predominantly of physical chemists. Lucas, in fact, remained the only true organic chemist on the staff for at least twenty years.[8] Lucas, if he were to talk chemistry at all, had to talk with the physical chemists. It is not surprising, then, that his research had a physical-chemistry orientation.

Lucas is well known for his books, including his 1935 text, *Or-*

ganic Chemistry, one of the first undergraduate texts on organic chemistry to introduce concepts of physical chemistry into the teaching of organic chemistry. His laboratory manual, written with David Pressman, is one of the classics in the field.[9] He is even better known for the Lucas reagent, used to distinguish primary, secondary, and tertiary alcohols, and for his research on the hydration of olefins, the stereochemistry of substitution in multiply substituted substrates (the beginning of neighboring group participation), aromatic substitution, and metal-pi bond interactions. Perhaps the most important "products" of his laboratory, at least as far as physical organic chemistry is concerned, were two of his graduate students, William G. Young and Saul Winstein.

By 1920 the stage was set for the next phase in the development of physical organic chemistry in the United States. Figure 2 shows the changes taking place at the four centers from 1920 to 1930.

At Harvard, James Bryant Conant joined the faculty in 1919, and before 1930 he was considered one of the most eminent organic chemists in the country.[10] He had been an undergraduate and a graduate student at Harvard, taking his Ph.D. degree jointly with Richards, the Nobel Prize–winning physical chemist, and Kohler, the organic chemist. This training made him a true physical organic chemist. His research was in physical chemistry, organic chemistry, and biochemistry. He worked on oxidation-reduction, electrochemistry, free radicals, acid strength of weak acids, and organic-reaction mechanisms using kinetics as an analytical tool. He also worked on complex organic syntheses. He taught courses in physical organic chemistry—though that name was not yet used—and in the chemistry of natural products. He claimed among his graduate students Louis Fieser, George Willard Wheland, Paul Bartlett, and Frank Westheimer.

Louis Hammett graduated from Harvard in 1916 and spent a year with Hermann Staudinger in Zurich, about a year in a World War I government laboratory, and a couple of years in the dye industry before entering graduate school at Columbia University in 1920. He took his Ph.D. degree with Hal Beans, a physical analytical chemist, in 1923 and then joined the Columbia faculty. He, too, was a true physical organic chemist (he memorized Kohler's course as an undergraduate). He had strong mathmatical leanings and was strongly influenced by the publications of Johannes Nicolaus Brønsted, Arthur Hantzsch, Gilbert Newton Lewis, and Alfred Werner, and by his association with Kohler, Nelson, and Beans. Hammett had a slow start publishing as a faculty member, but his 1928 paper, "The Theory of Acidity," was a strong indication of things to come.[11] Whereas Conant's career as a practicing physical organic chemist was almost over by 1930, Hammett's was just beginning. Hammett's best-known work, the development of the acidity function, H_o, and the structure-reactivity relationship known as the Hammett equation, was carried out and published in the 1930s, and his book *Physical Organic Chemistry* was published in 1940.

In the 1920s the chemistry program at the University of Chicago began to flounder. Stieglitz was growing old, was not doing much research, and had fallen into the bad habit of hiring his own students. In 1928 the program received a shot in the arm when Stieglitz hired Morris Kharasch, who had been an undergraduate and a graduate student at the University of Chicago.[12] He took a Ph.D. degree with Jean F. Piccard and also did some research with Stieglitz. After leaving Chicago, he went to the University of Maryland and was soon recognized as one of the important organic chemists in the country.

By the middle 1920s, Cal Tech was an important center for physical chemistry, and Linus Pauling characterized the direction and tenor of the department. Although Pauling was not an organic chemist—nor was he an inorganic chemist, a physical chemist, or a biochemist, yet, in a sense, he was all of these—he had an enormous impact on physical organic chemistry. Howard Lucas, who was beginning his research on olefins at Cal Tech, was blessed not only with brilliant young physical chemists but also with a young graduate student, William Young, from Colorado College by way of the Carnegie Institute. After finishing his Ph.D. degree in 1929, Young spent another year at the Carnegie Institute, this time as a National Research Council fellow. During this period he began the work on the allylic systems that was to be his major contribution to physical organic chemistry literature. He went to the University of California, Los Angeles, in 1930 and was instrumental in building a strong research program there. UCLA was Los Angeles State Normal School until it became part of the University of California in 1919. The program in the College of Letters and Science was initially a two-year program but was extended to four years in 1924. In 1933 a master's degree program was started, and a Ph.D. degree program was authorized in 1936.[13]

Figure 3 lists the major faculty changes taking place at the four centers from 1930 to 1940. In 1928, when Hammett was writing his first major paper, Kharasch was beginning to build an empire at the University of Chicago, Young was finishing up at Cal Tech, and Bartlett was beginning graduate work with Conant at Harvard. In 1931 Bartlett was a National Research Council fellow with Phoebus Aaron Levene at the Rockefeller Institute in New York. Having Bartlett work with Levene was Conant's attempt to interest Bartlett in biochemistry. During that year in New York, Bartlett worked nights and weekends at Columbia, where he had some contact with John Nelson and with the physical chemist Victor K. La Mer, but not with Hammett. After two years at the University of Minnesota, Bartlett was called back to Harvard to take over for Conant, who had left the community of chemists for the presidency of Harvard. Bartlett's research program at Harvard started slowly, but Kohler and Fieser sent him graduate students who were oriented toward physical chemistry and mathematics. Within a few years his reputation began to spread, and soon almost every physical organic chemist who ended up in a large research-oriented university had passed through Bartlett's laboratory as a graduate student or a postdoctoral fellow.

At the University of Chicago Kharasch was beginning his research program in free radicals, a monumental contribution.[14] But another of Kharasch's achievements, not so well documented in the literature, was building an outstanding staff of physical organic chemists in the late 1930s. After Stieglitz retired in 1933, the department was ruled by a three-man committee until 1945. Kharasch, as the organic chemist on the committee, had a great deal to say about the hiring of organic chemists.

In 1935 the department hired Weldon Brown, a physical chemist from Berkeley, to be an organic chemist. Hiring a physical chemist who, by his own admission, had almost no training in organic chemistry was one way of assuring that he was not "tainted" with classical organic chemistry. In 1936 the University of Chicago hired a former student of Kharasch, Frank Mayo, who had taken his Ph.D. degree in 1934 and spent a few years at DuPont, and the last graduate student of Conant, Frank Westheimer, who had been a National Research Council fellow with Hammett at Columbia. In 1937 the group was completed when George Willard Wheland was hired. Wheland had also been a graduate student of Conant and afterward worked as a postdoctoral associate with Pauling at Cal Tech and as a Guggenheim fellow with Christopher K. Ingold in London and Cyril Norman Hinshelwood at Oxford University. Herbert C. Brown, who had been a graduate student of Herman Schlesinger and a postdoctoral associate with Kharasch, was a member of the University of Chicago staff from 1939 to 1943. During this period he began his work on steric effects and basicity. In retrospect, this group was probably the largest cluster of first-rate physical organic chemists ever assembled at a single institution.

At Columbia, Hammett's work blossomed, but no significant personnel changes took place. Robert Elderfield was hired during the 1930s, but this addition had no impact on the physical organic chemistry program. Lucas remained the only organic chemist at Cal Tech until Carl Niemann was hired in the middle 1930s. Pauling and Wheland were busy developing resonance theory. At UCLA the research program was getting under way under the guidance of Young and Francis Blacet, a physical photochemist. Recognizing the shortcomings of organic chemistry at Berkeley, Young consciously set out to build a strong organic chemistry program at UCLA. Young's research program was helped considerably by the presence of a hardworking undergraduate, Saul Winstein. Winstein took his master's degree with Young, publishing seven papers in the process, but UCLA did not offer a Ph.D. program, so Winstein went to Cal Tech to work with Lucas. There he continued to turn out research, publishing another six papers before completing his degree requirements.

Figure 3 shows that the number of interconnections among the four centers was beginning to increase. The network was still limited, but its outlines can be discerned. Obviously, the work in England, particularly that of Ingold and Edward D. Hughes, was extremely important. The workers in the United States were certainly aware of that work and often found themselves competing with Ingold. However, in those days there were so many problems and so few workers that competition was hardly a necessity.

By 1941 several significant events had taken place. First, Hammett had published his book *Physical Organic Chemistry*, not the first book published in this area but one that seemed to have had a particularly profound effect.[15] It used for the first time the term *physical organic*, giving organic chemists interested in theory a corporate identity. They could now call themselves *physical organic chemists*. The book became a bible for organic chemists interested in mechanisms and structure-reactivity relationships, and it outlined a research program for the next generation of organic chemists.

Second, Winstein had finished at Cal Tech and gone to Harvard as a National Research Council fellow to work in Bartlett's laboratory. This association created a direct East–West pipeline that would serve as a conduit for many valuable exchanges.

Third, Robert Burns Woodward was settling in at Harvard and was quite close to Bartlett's group. He was not really a physical organic chemist, but he did more to change organic synthesis by introducing the ideas of organic theory than did any other single individual. Also, many of the outstanding physical organic chemists starting work in the 1950s and 1960s were Woodward's students.

And fourth, the community of chemists was gearing up for the war effort. World War II had a tremendous impact on physical organic chemists, giving them an opportunity to demonstrate that their form of organic chemistry could produce practical results. Before the war, and before Hammett's book, they were members of a subculture, looked upon with considerable suspicion. After the war, they were members of a well-respected community.

Figure 4 shows the changes that were taking place at the five centers from 1940 to 1950. UCLA was no longer an extension of Cal Tech, though close ties remained. At Harvard the only change was the coming of Woodward. Bartlett's research group was rapidly becoming *the* school of physical organic chemistry in the United States. Not until 1954, however, did a second physical organic chemist, Frank Westheimer, join the Harvard faculty.

At Columbia, Hammett's best research and writing had been done. From 1941 to the end of World War II, he was involved with the explosives laboratory at Bruceton, Pennsylvania. After the war he began a research program at Columbia again, but it was curtailed when he became department chairman. During the war, Arthur Cope was appointed to the Columbia staff, but because of his war work he was rarely at Columbia. Immediately after the war, he went to MIT.

The most significant change at Columbia was the hiring of William von E. Doering, who had been a graduate student of Reginald Patrick Linstead at Harvard. (Linstead was at Harvard only from 1939 to 1942. He went to England to work in the war effort and did not return to Harvard.) Doering went to Columbia after working with Woodward on the quinine synthesis and brought with him the Harvard approach to physical organic chemistry. Within a few years he would number among his students Donald Noyce, Andrew Streitwieser, Kenneth Wiberg, and Jerome Berson.

David Curtin spent five years at Columbia during the 1940s, and some of his best work in physical organic chemistry was done at that time. Curtin, Elderfield, and Doering left Columbia in the early 1950s. The resulting vacancies gave Hammett the opportunity to demonstrate his genius as an administrator. During his tenure as chairman, he put together an outstanding group of organic chemists that included Cheves Walling, Gilbert Stork, Ronald Breslow, and Thomas Katz.

Frank Mayo and Herbert Brown decided to leave the University of Chicago and subsequently became distinguished chemists. Wilbert Urry, who took his Ph.D. degree officially with Weldon G. Brown in 1946, was listed as a member of the faculty beginning in 1944. He continued the University of Chicago's tradition of research in organic mechanisms.

Cal Tech hired another organic chemist, but Lucas was still the only physical organic chemist on the staff. Significant changes were taking place at UCLA. After he left Harvard, Winstein spent a year at the Illinois Institute of Technology and then returned to UCLA for the rest of his career. In 1947 Donald Cram went to UCLA from a Ph.D. degree program with Fieser at Harvard. It is very likely that he surprised everyone by not turning out to be the natural-products chemist they thought they had hired. By 1949 it was clear that he would be a significant force in physical organic chemistry. Other organic chemists were on the staff at UCLA—Thomas Jacobs, Theodore Geissman, G. Ross Robertson—but Young, Winstein, and Cram were the nucleus of an extremely strong group of physical organic chemists. One of the major chemists coming from the UCLA program during this period was John D. Roberts.

Figure 4 clearly shows the increasing number of links between the base institutions. In the late 1940s other centers of physical organic chemistry were beginning to develop, including the three shown in Figure 5. The faculty members at these new centers had close ties to the original four centers.

At Berkeley, Noyce and Streitwieser became the nucleus of the group of physical organic chemists. Only two of a number of organic chemists hired at Berkeley between 1940 and 1955, both had been Doering's students at Columbia. Noyce went directly to Berkeley from Columbia, while Streitwieser did postdoctoral work at MIT with Roberts. At MIT, Cope was putting together a strong staff with a group of physical organic chemists who had come through Bartlett's laboratory. After Roberts finished his

Ph.D. degree with Young at UCLA, he spent a year as a post-doctoral fellow with Bartlett and then went to MIT in 1946. In 1953 he returned to the West Coast to take over the physical organic chemistry program at Cal Tech. C. Gardner Swain, one of Bartlett's Ph.D. students, also joined the MIT faculty in 1946, after a year at Cal Tech. In 1953 Fred Greene, another of Bartlett's Ph.D. students, went to MIT after a year as a postdoctoral fellow with Cram at UCLA.

At the University of Illinois, physical organic chemistry did not have an easy entry. Price, though one of Fieser's graduate students, had a physical-chemistry orientation and spent much of his short graduate career—he earned his Ph.D. degree in two years—talking with Bartlett.[16] He certainly tried to introduce an approach oriented more toward physical chemistry at the University of Illinois, and he was making some progress when he left for the University of Notre Dame in 1946. He was replaced at the University of Illinois by Elliot R. Alexander, who had been one of Cope's graduate students at Columbia and a postdoctoral fellow with Bartlett. Alexander would have had a strong impact on physical organic chemistry had he not died in a tragic plane crash in 1951. He was replaced at the University of Illinois by Curtin, who had been at Columbia. At the same time, programs in physical organic chemistry were starting at the University of Wisconsin, Iowa State University, and a number of other universities considered the last bastions of classical organic chemistry.

By way of proclaiming their self-recognition as a community, the physical organic chemists organized the first Organic Reaction Mechanisms Conference in 1946. The conference, held at Notre Dame, was organized by Price and Bartlett. Herbert C. Brown's work on steric effects was already recognized as important, but perhaps his title, "Non-Classical Steric Effects," was a harbinger of things to come.[17] In less than thirty years, the workers in physical organic chemistry in the United States had grown from a few independent souls who were willing to challenge the established authority of classical organic chemistry to a spreading, but closely knit, network that could and would call itself a community.

NOTES

1. This history of physical organic chemists has been examined and reported previously by the author in a series of papers presented at the 13th, 14th, and 17th Middle Atlantic regional meetings, and the 178th, 182nd, and 184th national meetings, of the American Chemical Society.

2. James B. Conant, "Elmer Peter Kohler," *Biographical Memoirs* (*National Academy of Sciences of the United States of America*), 27 (1952), 265–91, and *My Several Lives* (New York, 1970); Louis F. Fieser, "Arthur Michael: August 2, 1853–February 8, 1942," *Biographical Memoirs* (*National Academy of Sciences of the United States of America*), 46 (1975), 331–66; D. S. Tarbell, "Organic Chemistry: The Past 100 Years," *Chemical and Engineering News*, April 6, 1976, pp. 111–12.

3. *American Men of Science* (9th ed.; New York, 1955); New York *Times*, October 11, 1939, p. 29, March 9, 1940, p. 32, and November 16, 1965, p. 43; interview with Hammett, May 1, 1978, conducted by author, on deposit at the American Institute of Physics Center for History of Physics, New York.

4. M. L. Wolfrom, "John Ulric Nef, 1862–1915," *Biographical Memoirs* (*National Academy of Sciences of the United States of America*), 34 (1960), 204–27; W. A. Noyes, "Julius Stieglitz, 1867–1937," *ibid.*, 21 (1939), 275; Herbert N. McCoy, "Julius Stieglitz, 1867–1937: A Biographical Sketch," *Journal of the American Chemical Society Obituaries*, 60 (1938), 3–18; D. S. Tarbell, "Solomon F. Acree," paper presented at the 182nd National Meeting of the American Chemical Society, New York, August, 1981.

5. *Annual Register of the University of Chicago*, July, 1906–July 1907, and July, 1907–July, 1908.

6. Robert F. Kohler, Jr., "The Origin of G. N. Lewis's Theory of the Shared Pair Bond," *Historical Studies in the Physical Sciences*, 3 (1971), 343; John W. Servos, "G. N. Lewis: The Disciplinary Setting," *Journal of Chemical Education*, 61 (1984), 5–10; Melvin Calvin, "Gilbert Newton Lewis: His Influence on Physical-Organic Chemists at Berkeley," *Journal of Chemical Education*, 61 (1984), 14–18; G. E. K. Branch and Melvin Calvin, *The Theory of Organic Chemistry* (New York, 1941).

7. William G. Young and Saul Winstein, "Howard Johnson Lucas," *Biographical Memoirs* (*National Academy of Sciences of the United States of America*), 43 (1973), 163–76.

8. John Servos, "The Knowledge Corporation: A. A. Noyes and Chemistry at Cal-Tech, 1915–1930," *Ambix*, 23 (1976), 175.

9. Howard J. Lucas, *Organic Chemistry* (New York, 1935); Howard J. Lucas and David Pressman, *Principles and Practice in Organic Chemistry* (New York, 1949).

10. Martin Saltzman, "James Bryant Conant and the Development of Physical Organic Chemistry," *Journal of Chemical Education*, 49 (1972), 411–12; Paul D. Bartlett, "James Bryant Conant," *Biographical Memoirs* (*National Academy of Sciences of the United States of America*), 54 (1983), 91–124; G. B. Kistiakowsky and F. W. Westheimer, "James Bryant Conant, 26 March 1893– 11 February 1978," *Biographical Memoirs of Fellows of the Royal Society*, 25 (1979), 209.

11. Autobiographical notes of Louis P. Hammett, deposited in Louis P. Hammett file, American Institute of Physics, New York; interview of Hammett by author; Louis P. Hammett, "The Theory of Acidity," *Journal of the American Chemical Society*, 50 (1928), 2666–73.
12. F. W. Westheimer, "Morris Selig Kharasch," *Biographical Memoirs* (*National Academy of Sciences of the United States of America*), 34 (1960), 123–52; "Morris S. Kharasch, 1895–1957," Memorial Service, University of Chicago, January 14, 1958.
13. *American Men and Women of Science* (13th ed.; New York, 1976), 5002; Verne A. Stadtman (ed.), *The Centennial Record of The University of California* (Berkeley, 1967).
14. Cheves Walling, "Forty Years of Free Radicals," *Organic Free Radicals*, ACS Symposium Series, 69 (Washington, D.C., 1978), 3–11.
15. Louis P. Hammet, *Physical Organic Chemistry* (New York, 1940). Other advanced organic chemistry textbooks with a physical-chemistry orientation published in the late 1930s and early 1940s were: William A. Waters, *Physical Aspects of Organic Chemistry* (London, 1935), H. B. Watson, *Modern Theories of Organic Chemistry* (New York, 1937), Branch and Calvin, *The Theory of Organic Chemistry*, and A. Edward Remick, *Electronic Interpretations of Organic Chemistry* (New York, 1943).
16. Inteview with Price, May, 1979, conducted by author, on deposit at the Center for History of Chemistry, Philadelphia.
17. The other papers presented at the 1946 conference were three on displacement reactions: neighboring group participation, by Winstein, Price, and Bartlett; two on free radicals, by Urry and Mayo, one on nitration, by Westheimer; one on resonance and reactivity, by Wheland; and one on entropy and reactivity, by Hammett.

The Familiar and the Systematic:
A Century of Contention in Organic
Chemical Nomenclature

Students beginning a study of organic chemistry learn that there are officially approved ways of naming the many individual compounds, and they may even learn that the author of their textbook misdirects them in connection with such names. They are less likely, however, to learn as much history of the official nomenclature as of the reactions and theories included in the course. Yet a look at the history of official nomenclature provides some interesting insights into how chemists go about their work, how emphases and influences shift during the development of a field. Perhaps surprisingly, the characteristic of chemists that seems to persist through the history of organic nomenclature is their resistance to change—this in a group whose excitement is often associated with logical trains of thought and new reaction schemes and theoretical concepts.

Names of compounds are now based on structure, but names were coined before structures were known or even acknowledged. In a landmark paper in 1832, Justus Liebig and Friedrich Wöhler used *benzoyl* as the name for the molecular fragment that persisted in a series of reactions.[1] The name was not associated with structure, but just with the C_7H_5O (then $C_{14}H_{10}O_2$) fragment; it continues in official use today for the same fragment and now also for a particular structure. The need for names always outruns the prescribing of rules for names. Some of the first-formed names like benzoyl found such wide acceptance and use that systemization, when it came, had to accommodate them. Frequently, these early, persistent names, such as formic acid (Latin *formica*, or "ant"), reflected a first or significant source of the compound. Similar practice continues today, especially with natural products of unknown structure.

Some chemists in England showed an early concern for good nomenclature practice. In 1866, in a long footnote, August Wilhelm von Hofmann proposed systematic names for hydrocarbons and used a sequence of suffixes, following the order of the vowels (*-ane, -ene, -ine, -one, -une*), to signal increasing degrees of unsaturation. A decade later, Henry E. Armstrong, a member of the Committee on Publication of the *Journal of the Chemical Society*, inserted in the 1876 volume of the journal a footnote to a lengthy abstract of a paper by Wilhelm Körner, changing Körner's use of *ortho-, meta-,* and *para-* designations from the system in favor in Germany to the more widely accepted one still in use. Later the same year, the journal's list of the Chemical Society's monthly meetings records that Armstrong presented a paper, "On Systematic Nomenclature," at the April 6, 1876, meeting.[2] The discussion on that paper carried over to the next meeting on April 20, but the record does not reveal any information about the contents of the paper or the discussion.

The *Journal of the Chemical Society* published lengthy abstracts of chemistry papers that had appeared in other journals, and the 1879 volume carried an item entitled "Instructions to Abstractors." In addition to supporting initiative by abstractors such as that displayed by Armstrong in his footnote to the Körner abstract, the instructions included some rules about names of compounds that were an approach to standardization of nomenclature. Some of these rules, however, did not accord with the rules adopted thirteen years later by the commission at Geneva. Two of the earlier rules stated, "The isomeric hydrocarbons are usually most conveniently represented by names indicating their relation to methane: for example, $CH_3 \cdot CH_2 \cdot CH_2 \cdot CH_3$ = propylmethane," and "Term the hydrocarbons C_2H_4 and C_2H_2 ethylene and acetylene respectively (not ethene and ethine)." Among the rules that clearly anticipated the Geneva commission's systematic approach, one advised, "Distinguish all alcohols, *i.e.,* hydroxyl-derivatives of hydrocarbons, by names ending in *ol, e.g.* quinol . . . glycerol . . . instead of hydroquinone . . . glycerin." "Instructions to Abstractors" was clear on one matter that presaged a position taken by *Chemical Abstracts* nearly one hundred years later, to the consternation of many chemists: the use of standard nomenclature in an abstract regardless of the names used by the original authors. (The instructions contain the interesting spelling *radicles*.) The final two sentences of the publication are instructive about the general state of chemical nomenclature in 1897 and about the position of the journal's editor: "The compounds of basic substances with hydrogen chloride, bromide, or iodide, should always receive names ending in *ide* and not *ate*, as morphine hydrochloride and not morphine hydrochlorate.—*The Editor's decision in all matters connected with the Abstracts must be considered final.*"[3]

By 1889 organic chemical nomenclature problems were sufficiently troublesome that the International Chemical Congress, meeting in Paris that summer, appointed a special section, later identified as a commission, "to consider the unification of chemical nomenclature." The members of the commission living in Paris took on the responsibility of preparing a set of recommendations that became the focus of a conference in Geneva that convened on Easter Monday, 1892. Besides the official record, reports of that conference and its resolutions were published in *Berichte der Deutschen Chemischen Gesellschaft* by Ferdinand Tiemann and in *Nature* by Armstrong, both members in attendance at the conference.[4]

Journal space was surely not at such a premium as it is today, for the reports include fascinating details that would not appear in current journal reports of such a meeting. In Armstrong's report, which has much of the flavor of an informative letter written when letter writing was still a cultivated art form, we find such disclosures as, "It is worth mentioning, as an illustration of the sympathetic treatment accorded by public bodies in France to men of science, that the Paris-Lyons-Marseilles Railway Company granted a reduction of one half on the fare over their line to members of the Congress." (That discount surpasses any that the American Chemical Society has been able to arrange.) After a first day of morning and afternoon sessions, Armstrong reported, "We met in like manner on the two following days, and the final sitting took place on Friday morning, but many had left before this."[5] The pattern of meeting attendance seems to have been set at the Geneva conference at least as much as the pattern of nomenclature.

The Geneva conference was attended by thirty-four prominent chemists from nine European countries; half of the group were from France and Switzerland, and the others were from Germany, England, Italy, Austria, Belgium, Romania, and the Netherlands. (Actually, the city rather than the country from which each conference member came was named in the official record. It is interesting that the two reports, in *Berichte* and in *Nature*, are not in complete accord in identifying the cities.)

In the opening session Stanislao Cannizzaro, from Rome, whose influence on the first international chemical conference in Karlsruhe thirty-one years before had been most lasting, showed his influence again. He moved successfully that the complimentary but rather inefficient prevailing practice—followed at Karlsruhe—of having a different chairman each day be abandoned in favor of having a single chairman. Charles Friedel, who had chaired all the meetings of the committee in Paris, was chosen permanent chairman of the conference.

The group worked diligently and cooperatively. The apprehension that some had brought to the conference about even the possibility of devising a suitable scheme rather quickly vanished as all joined in the work on the draft developed by the committee in Paris. Forty-six resolutions were adopted. The intent was to prescribe a single name for each substance and also to have as few departures from popular usage as possible. The longest continuous chain as parent was firmly emphasized, and unbranched saturated hydrocarbons were to be named without any prefix. Thus, the prefix *n-* designating an unbranched isomer has never had any official standing and was omitted as unnecessary even in the rules from the first conference. Yet that prefix keeps showing up today in textbooks and journals.

Cycloalkane type names were adopted in preference to polymethylene type names, but hardly any other attention was given to cyclic compounds. Some systematic endings (*-ene, -ine, -ol, -al, -one, -oic acid*) were adopted; some nonsystematic, familiar names were retained (for example, the names of the first four unbranched alkanes), and some abandoned (for example, *mercaptan*); some issues were left undecided (for example, nitriles). The primary recommendation of the conference was that endings should be given to all names to indicate the class of the compound, but the saturated hydrocarbon parent was paramount. "The numbering of the hydrocarbons is conserved for all their products of substitution."[6]

The members of the conference aspired to a definitive document, but they were realistic enough to recognize that their work, inspired though it was, was not complete. Armstrong wrote in his report, "The resolutions passed at the meetings . . . are in no way to be taken as in all respects final, but they will serve to prepare the way and to indicate the lines on which the work is to be carried out." Perhaps reflecting his own longtime involvement in promoting good nomenclature practice as much as a consensus, Armstrong looked toward the future and wrote, "The value of such systematic nomenclature to original workers as well as to students cannot be over-estimated, and few who are qualified to take part in such work will grudge the time they spend on it."[7]

The resolutions of the Geneva conference did not receive complete acceptance. In fact, opposition to systematic, structure-based names was lively in some quarters. For example, the president of the Chemical Society, A. Crum Brown, spoke rather forcefully against the idea of systematic nomenclature after hearing a paper on nomenclature of cyclic compounds by Armstrong, who himself did not like cycloalkane type names.[8] But the resolutions received a boost when they were immediately adopted and then used as the basis for the names listed in those volumes of the third edition of Friedrich Konrad Beilstein's *Handbuch der Organischen Chemie* that were published in the late 1890s. Like Ira Remsen of the United States, Beilstein had been a member of the original commission but was not a member of the Geneva conference.

In spite of the acknowledgment by the members of the 1892 conference that the work was incomplete and Armstrong's faith in the willingness of chemists to undertake that work without grudging, no further official nomenclature group was convened until the 1920s. The increase in the number of compounds and the use of "obviously incorrect" names for many of the newer ones, perhaps because of incomplete guidance by the Geneva conference rules, led the International Union of Chemistry (IUC) in 1922 to appoint a working committee comprising the editors of *Chemical Abstracts*, the *Journal of the Chemical Society*, and *Le Bulletin de la Société Chimique de France*. The committee was subsequently enlarged to include representatives of *Gazzetta chimica*, *Helvetica Chimica Acta*, and *Recueil des Travaux Chimiques des Pays-Bas*.[9] Two among this group, the members from Italy and Switzerland, had also been members of the Geneva conference. Because of the superlativeness of German chemistry at the time, the omission of a German editor is surprising. Only at the final meeting of the working committee in 1930 was a German representative, the editor of *Beilstein*, present.

Some members of the Geneva conference had been members of the publication committees of a few journals, but none were identified as editors. The IUC committee members were all editors, specifically chosen because of that role. In effect, the responsibility for settling nomenclature matters had been passed from the principal research investigators to the journal and abstract editors. Yet, intriguingly, the Definitive Report finally adopted by the committee and by the IUC in Liége in 1930 included in its introduction the statement, "An official nomenclature for indexes which was desirable 40 years ago has today become unnecessary"—because of the new edition of *Beilstein's Handbuch der Organischen Chemie* and the use of tables or formulas in journals and the major abstracts publications.[10]

The new official rules differed in two substantial respects from the 1892 Geneva conference rules: 1) Flexibility in names was allowed rather than insistence on a single name, and 2) the functional group became the principal focus for numbering the parent chain. Other modifications included the replacement of *-ine* as the ending for the C≡C (an ending that never caught on because of confusion with amines) by *-yne* and the designation only of the principal functional group by a suffix. The continuing confrontation between the systematic and the familiar is illustrated in rule 59: "Univalent radicals derived from aromatic hydrocarbons by loss of an atom of hydrogen from a ring will in principle be named by changing the termination ene to yl. [By this rule, C_6H_5 would be benzyl.] However, the radicals C_6H_5 and $C_6H_5 \cdot CH_2$ will continue provisionally to be named *phenyl* and *benzyl* respectively. Moreover, certain abbreviations sanctioned by custom are authorized, such as *naphthyl* in place of *naphthalyl*." In all, sixty-seven rules were set forth, some dealing with issues not included in the Geneva conference resolutions. Entry 68 was not a rule but a promise that a catalog of cyclic systems and their numbering, based on proposals by the *Chemical Abstracts* representative, Austin M. Patterson, was in preparation by the National Research Council and the American Chemical Society. (Patterson's report in the *Journal of the American Chemical Society* included the added comment in small type, "Work on this project is now in abeyance pending better financial conditions for publication.")[11]

At the Geneva conference the United States had not even been represented, but in the IUC committee its representative, the editor of *Chemical Abstracts*, had a dominating influence. The official written language of both groups was French, and the English version of the 1930 document was identified as a translation in the *Journal of the American Chemical Society*, but not in the *Journal of the Chemical Society*. An amusing, inconsequential difference between the two English versions appears in the final sentence: "In order to avoid any confusion, the Committee recommends that a scheme of numbering be given at the foot of each memoir" (*Journal of the Chemical Society*) and "In order to avoid all confusion the Commission recommends placing a scheme of numbering at the head of each article" (*Journal of the American Chemical Society*).[12] The words in the French version designating the place for the scheme are *au dessus*.

The major intent of the committee that generated the 1930 Liége rules was probably captured in Patterson's added comment at the beginning of his report of the rules: "The Committee members were apparently agreed that any attempt to introduce sweeping changes in terminology, however consistent they might be, would prove impractical and undesirable in the present stage of the science. The purpose of the rules, therefore, is rather to unify existing practice as far as possible by eliminating objectionable names and guiding future naming along desirable lines. The proposal (in Geneva rule 1) of an official name for each organic compound has been abandoned." Rule 13 permitted *phene* for C_6H_6, but Patterson commented, "'Phene,' while logically related to 'phenol,' is not likely to replace 'benzene.'"[13] (*Phene* was not even mentioned as an alternative in the 1957 rules.) Similarly, from the other side, general application of rule 20 would have named C_6H_5-OH *benzenol,* but *phenol* was specifically retained. Familiarity continued to win over systematics. Some of the 1930 rules left ambiguities, even for committee members, as revealed by the comments Patterson interspersed throughout his report of the rules.

The 1930 Liége rules were supplemented by shorter reports in the later 1930s, but by 1947, the International Union of Pure and Applied Chemistry (IUPAC), the organization earlier identified as the IUC, decided that organic chemical nomenclature again required comprehensive attention. This time both editors and research investigators of note who were not editors were involved in the work of the commission. Progress reports were published in the official records of the IUPAC conferences, and a new, much longer set of rules, identified as the 1957 IUPAC Rules, was adopted in 1957. Acknowledging the extensive growth of nomenclature, and nomenclature problems, since the last definitive report, the commission "confined its efforts to codifying sound practices which already existed, rather than to originating new nomenclature," but hinted that new nomenclature might occupy the commission later.[14]

While still retaining flexibility in acceptable nomenclature styles, the 1957 document seemed to shift emphasis back slightly toward the 1892 Geneva conference's position of one acceptable name for each compound. "The Commission hopes that elimination of alternatives may become acceptable as the merits of one method become more generally recognized."[15] Not yet quite

bold enough to say forthrightly which method should become
more generally recognized, the commission at least gave a reasonably strong indication that merit was not equally shared
among all methods. English, not French, was now very clearly
the first language of the rules, and the style of publication had
changed so that none of the side disclosures that had enlivened
Armstrong's account of the Geneva conference were ever included again. Comments inserted in the publication of the 1957
rules in the *Journal of the American Chemical Society* highlighted
differences between *Chemical Abstracts* index practice and the
IUPAC rules, but *Chemical Abstracts* practice had been significantly influenced in the commission's deliberations and
decisions.

The Commission on Nomenclature of Organic Chemistry has
continued in active status, revising and expanding the rules on
a continuing basis. Proposals from investigators in new or expanding areas of organic chemistry have received active consideration. In 1960 IUPAC began issuing *Pure and Applied Chemistry*
as the official journal of the International Union of Pure and Applied Chemistry. Reports of the several IUPAC commissions, including the Commission on Nomenclature of Organic Chemistry, have since been published in that journal, along with papers
from IUPAC conferences and symposia. The 1957 rules were subdivided into Sections A (Hydrocarbons) and B (Fundamental
Heterocyclic Systems). Section C (Characteristic Groups Containing Carbon, Hydrogen, Oxygen, Nitrogen, Halogen, Sulfur,
Selenium, and/or Tellurium) was published in 1965, and a second edition of Sections A and B, containing some corrections,
was published the following year.[16]

In 1969 a new (third) edition of Sections A and B was adopted,
reflecting firmer stands by the commission on some matters.[17]
Whereas earlier editions had permitted citations of substituent
side chains either in order of complexity (specified first in the
rule) or in alphabetical order, the 1969 rules specified only alphabetical order. Similar firmness emerged in rules for heterocyclic systems.

Responding to the expansion of organic chemistry itself as

well as to initiatives of some research investigators concerned about the need for new nomenclature, the commission extended its rule making into new areas, and in 1979 a new volume with four new sections was published. Sections A, B, and C were the same as in 1969, except that they were "corrected for material errors."[18] Section D (Organic Compounds Containing Elements That Are Not Exclusively Carbon, Hydrogen, Oxygen, Nitrogen, Halogen, Sulfur, Selenium, and Tellurium) was issued jointly by the organic and the inorganic nomenclature commissions. Section E (Stereochemistry) was based on 1974 recommendations incorporating the E,Z and R,S specifications of configuration. Section F (General Principles for the Naming of Natural Products and Related Compounds) did not actually have final approval by IUPAC, but Section H (Isotopically Modified Compounds) did, just preceding the 1979 edition. A later edition reorganizing some of the material from one section to another was promised in the introduction.

The page and a half of 46 resolutions adopted by the 1892 Geneva conference had grown in 87 years to a book of 559 pages. And the work of the commission was still not complete, nor will it ever be so long as new organic chemistry is happening. From the beginning of efforts toward standardization of organic nomenclature, a name simply for identification has not been enough. The name should also connote the chemical behavior to be expected of the compound. As the 1979 *Nomenclature of Organic Chemistry* says in Section D, "The concept of a 'derivative' is peculiar to organic chemistry."[19] Contrast, for example, the names of $PO(OCH_3)_3$ formed on the basis of organic chemistry rules and inorganic chemistry rules: *trimethyl phosphate* and *trimethoxooxophosphorus*, respectively.

Chemical Abstracts Service has been guided by, and has provided significant guidance for, IUPAC nomenclature rules, but it has not been bound by those rules. With its focus on indexing and retrieving chemical information, Chemical Abstracts Sevice has sometimes deliberately chosen a name alternative to IUPAC names for use in *Chemical Abstracts* indexes. Chemical Abstracts Service has been more willing, perhaps more driven, than IUPAC to be firm in using a single name for a compound. Usually, that single name is among those permitted by IUPAC rules, but sometimes it is not the one in most common use.

In 1972 Chemical Abstracts Service began to use significantly revised index names for compounds. While firmly emphasizing a single index name for each compound, Chemical Abstracts Service acknowledged in its *Index Guide* that "the use of this invariant index name for citation throughout every context in the scientific community is neither practicable nor desirable." Even so, by 1975 grumblings of dissatisfaction from some users of *Chemical Abstracts* found expression in earnest, almost amusing, letters that were circulated and published.[20] Some prominent organic chemists were anguished, for example, that toluene was no longer indexed as *toluene* but as *methylbenzene* (actually, *Benzene, methyl-*), right along with other substituted benzenes (*Benzene, substituent-*). Other correspondents, however, cited examples in which the retrieval of literature was facilitated by the *Chemical Abstracts* approach to indexing.[21] Familiarity and systematization were still antagonists. But the rapid expansion of on-line literature retrieval has strongly tilted the contention in favor of the systematic names. The emotional pinch on chemists will disappear only if the systematic becomes the familiar.

To what conclusions or position does this retrospective on organic chemical nomenclature bring us? Faultless communication of chemical identity is worthy of encouragement but is more difficult than many chemists are willing to concede.[22] The increasing shift toward the exclusive use of good, systematic nomenclature in catalogs of the suppliers of chemicals may be as effective as any teaching or preaching to bring about the identity of the official and the familiar. Teachers and textbook authors, however, continue to have responsibilities in this matter, which are too often neglected, even flouted. Each class of students being introduced to organic chemistry is unencumbered with unapproved, unsystematic names and thus provides a fresh beginning for the teacher. We teachers readily spare them the burden of old, unsatisfactory syntheses that we had to learn in favor of teaching new transformations that really do work. On the other hand, some names that we teachers and authors had to learn are apparently more difficult to leave behind, however unsatisfactory they are for good communication. The change-over can be made successfully, and textbook authors actually have a major role in making the unsatisfactory also the unfamiliar. When we do a good job of teaching organic nomenclature, by word and example, good nomenclature practice becomes so familiar to our students that, unaware of the old antagonisms, they will not even thank us for so advantaging them.

As the saying goes, familiarity breeds. With organic chemical nomenclature, let us try to make the offspring impeccancy. Life after publishing will be so much easier for all of us.

NOTES

1. Friedrich Wöhler and Justus Liebig, "Untersuchungen über das Radikal der Benzoesäuer," *Annalen der Pharmacie*, 3 (1832), 249–82.
2. A. W. Hofmann, "On the Action of Trichloride of Phosphorus on the Salts of the Aromatic Monamines," *Proceedings of the Royal Society* (London), 15 (1866), 57–58; H. E. Armstrong, *Journal of the Chemical Society*, n.s., 1 (1876), 207; "Proceedings of the Chemical Society," *Journal of the Chemical Society*, n.s., 2 (1876), 685.
3. H. Watts, "Instructions to Abstractors," *Journal of the Chemical Society*, 35 Transactions (1879), 278, 280, 281.
4. Ferdinand Tiemann, "Ueber die Beschlüsse des internationalen, in Genf vom 19. bis 22. April 1892 versammelten Congresses zur Regelung der chemischen Nomenclatur," *Berichte der Deutschen Chemischen Gesellschaft*, 26 (1893), 1595–1631; Henry E. Armstrong, "The International Conference on Chemical Nomenclature," *Nature*, 46 (1892), 56–59.
5. Armstrong, "The International Conference on Chemical Nomenclature," 56.
6. *Ibid.*, 58.
7. *Ibid.*, 57.
8. *Proceedings of the Chemical Society* No. 114 (1892), 127–31.
9. "Definitive Report of the Commitee for the Reform of Nomenclature in Organic Chemistry," *Journal of the Chemical Society*, (1931), 1608.
10. *Ibid.*, 1607.
11. *Ibid.*, 1615; Austin M. Patterson, "Definitive Report of the Commission on the Reform of the Nomenclature of Organic Chemistry, Translation with Comments," *Journal of the American Chemical Society*, 55 (1933), 3925.
12. "Definitive Report of the Committee for the Reform of Nomenclature," 1616; Patterson, "Definitive Report of the Commission," 3925.
13. Patterson, "Definitive Report of the Commission," 3907–3908, 3911.
14. "Definitive Rules for Nomenclature for Organic Chemistry," *Journal of the American Chemical Society*, 82 (1960), 5545.
15. *Ibid.*
16. International Union of Pure and Applied Chemistry, *Nomenclature of Organic*

Chemistry: Definitive Rules for Section C (London, 1965); *ibid., Nomenclature of Organic Chemistry: Definitive Rules for Section A . . . Section B . . . Steroids* (London, 1966).

17. *Ibid., Nomenclature of Organic Chemistry: Sections A, B, and C* (3rd ed.; London, 1971).

18. *Ibid., Nomenclature of Organic Chemistry: Sections A, B, C, D, E, F and H* (Oxford, 1979), xv.

19. *Ibid.*, 399.

20. Norman Donaldson *et al., "Chemical Abstracts* Index Names for Chemical Substances in the Ninth Collective Period (1972–1976)," *Journal of Chemical Documentation,* 14 (1974), 3–15; *Chemical Abstracts Index Guide, 1972–1976* (Columbus, Ohio), Appendix 4, p. 118I; Dennis C. Owsley *et al.,* Michael Dub, "*CA* Nomenclature," *Chemical and Engineering News,* May 5, 1975, p. 3; Theodore O. Groeger, "*CA* Nomenclature," *Chemical and Engineering News,* August 18, 1975, p. 3.

21. Russell J. Rowlett, Jr., "*CA* Nomenclature," *Chemical and Engineering News,* May 5, 1975, pp. 3, 46–47; Jean S. Peterson, *ibid.,* June 16, 1975, pp. 3, 53.

22. Two brief commentaries on some of the general philosophy and difficulties of organic nomenclature are strongly recommended: Charles D. Hurd, "The General Philosophy of Organic Nomenclature," *Journal of Chemical Education,* 38 (1961), 43–47, and John H. Fletcher, Otis C. Dermer, and Robert B. Fox, "Common Errors and Poor Practices," in *Nomenclature of Organic Compounds: Principles and Practice* (Washington, D.C., 1974), 120–29.

Chemage: A Compendium of Chemical Trivia

SOME years ago, a visitor asked my eldest child, then about four years old, what his daddy did. My son replied that his daddy "did chemage," and everyone was appropriately amused. Thinking it over afterwards, I decided that the word, which seems to be a combination of *chemistry* and *garbage*, was particularly appropriate to designate the collection of items, most clipped from the original literature, that I have accumulated for perhaps the past thirty years.

There are valid reasons for collecting chemage: It tells something about what chemists study, something about the most and the least of the field, and mostly, something about the maturity of the field. Much of what I shall present would have been publicly unacceptable forty years ago. We chemists were far too concerned with our professional dignity at that time, reflecting a basic insecurity in our public image. We are now much less concerned about being caught in a stained lab coat and far more willing to air publicly our foibles and amusements. I prophesy that we shall see increased amounts of chemage in the future.

Each newly recruited chemist, on first encountering the languages of chemistry, both written and structural, finds it almost irresistible to play games with the new toys. There exists a large, and still growing, number of punny structures, many of which have been around for more than two generations.

ORTHODOX   METAPHORS   PARADISE

127

These designations can evolve, just as the word *square*, popular in the forties, later became *cube* to designate a person who was uninformed about the latest in adolescent interests.

MD
7-UP
UNORTHODOX

One can make topical references (which quickly become out-of-date), for example, the spiro structure below:

A   E   SPIRO
N   AGNEW
G   W

Many variations exploit geometrical isomerism around the ethylene bond.

MA   H
C=C
H   PA
TRANSPARENCE

H   H
C=C
Boom   Bah
CIS-BOOM-BAH

Two of the earliest punning formulas were:

HIOAg   RIN(Sn)$_2$
HI YO, SILVER   RIN-TIN-TIN

Two newer ones are:

MERCEDES   OUT-ON-PYRROLE
BENZENE

My file contains a great many more, and my students show impressive creativity in developing new ones.

There are other possibilities that involve the use of words rather than structures (though these tend to be somewhat less interesting). A typical example is, "For what class of compounds is Paul Revere best known?" The answer: "nitrides."

The metric system proves to be a natural for word play.

1 x 10^{-12} doors = 1 picadoor
1 x 10^1 cards = a dekacards
1 x 10^{-2} claus = a centiclaus

The prefixes themselves can be utilized.

A millipicture is a picture worth one word.

There are the peculiarly balanced equations

$$1 \text{ Ba} + 2 \text{ Na} \rightarrow 1 \text{ Banana}$$
$$4 \text{ Nd} + P_4 \rightarrow 4 \text{ Pd} + 2 \text{ N}_2$$

and jokes based on the periodic table.

Physician says: "I *He*."
"It's none of your *Bi* how much *Fe*."

One of the difficulties in the classroom is communicating the awesome magnitude of some of the exponential numbers that chemists employ. Here are two examples that occasionally help:

One *ppm* corresponds to one jigger of vermouth in a tank-car load of gin. (A *very* dry martini!)
One *ppb* corresponds to one step toward the moon.

Many of us have a fundamental fascination for reports of "the most" and "the least"—all the extremes of everyday events, including areas involving professional activities. *Chemical Abstracts* is a source of a great many of them for which there is insufficient space here. The following questions, all of which can be answered with a reference to the original professional literature, constitute a "mini"-Guinness book of records, perhaps uniquely of interest to chemists.[1]

1) What is the largest number of authors of a single paper? 142.
2) What is the longest laboratory reaction time reported? 7 years.
3) What is the largest number of structures (all isomers!) on a single page? 412.
4) What is the common title of the most papers in a numbered, continuous series? "Studies on the Syntheses of Heterocyclic Compounds," written by Tetsuji Kametani *et al.* Paper 460 was published in 1972 and paper 715 in 1977, both in the *Journal of Organic Chemistry.* (This record reflects an average of a paper per week for the approximately 250-week period.) Kametami has by now exceeded paper 1000 in this series.
5) What is the number and the name of the most authors with the same surname on a single paper? Four, Patel. (A fifth Patel is mentioned in an acknowledgment, and the paper comes from a university with the same name as the authors.)
6) Has the same paper ever been published twice in the same journal? Yes.

7) How long is the shortest publication? One sentence: "The lowest field doublet in the 64.16 Mc./sec. ^{11}B n.m.r. spectrum of decaborane is converted into a singlet in the spectrum of a diethyl ether solution of 1,2,3,4-tetradeuterodecaborane, thus allowing conclusive assignment of this resonance to the 1(3) borons." There are other one-sentence papers; this is the shortest.

At some time or other, most of us—whether from simple fatigue, an attack of whimsey, or perhaps a dare—have entertained the notion of carrying out some professional activity in an unorthodox fashion. Getting something past an eagle-eyed editor is a goal of long standing. One of the most successful attempts appears in an article published in the *Journal of the American Chemical Society* in 1955.[2] By looking very carefully—possibly requiring a low-power magnifying glass—at the upper middle portion of a moderately complex block diagram (Fig. 1 in the article), one will observe a tiny stick figure fishing in container A. What is more, the fisherman is about to catch a fish!

Occasionally, authors have employed unorthodox forms of presentation. Those readers familiar with Longfellow's epic poem *Hiawatha* will recognize its meter in the report in the *Journal of Organic Chemistry* on some of Joseph F. Bunnett's benzyne research. The opening paragraph reads:

Reactions of potassium amide
With halobenzenes in ammonia
Via benzyne intermediates occur.
Bergstrom and associates did report,
Based on two-component competition runs,
Bromobenzene the fastest to react,
By iodobenzene closely followed,
The chloro compound lagging far behind,
And fluorobenzene to be quite inert
At reflux ($-33°$).
Reactions with *para*-dihalobenzenes,
In which the halogens were not the same,
The same order of mobility revealed,
But differences in reactivity
Were somewhat less in magnitude.

The journal's editor commented in footnote two of the paper, "Although we are open to new styles and formats for scientific publication . . . manuscripts in this format face an uncertain future."[3] For the most effective results, one should read the paper aloud with the style of elocution of two or more generations ago.

Perhaps the most unorthodox presentation was by the physician, Dr. Howard Shapiro, who *sang* his paper at a national meeting while accompanying himself on a guitar. He used the "talking blues" format most skillfully, as is apparent in these few stanzas.

> Now Britton Chance and all his buddies
> Have shown, in their extensive studies,
> That PO_2 shifts 'cross the range
> Give a twenty percent fluorescence change
> They call what they're measuring redox state . . .
> There's been a lot of concentration
> On means of organ preservation,
> But, though we have been pretty clever
> We don't keep them alive for ever.
> A few days, maybe . . . next slide, please . . .
> The transplant team finds nought so irking
> As organs that abandon working
> Between the time they leave the donor
> And when they're placed in their new owner.

And if that were not enough, he published a report on fluorescent dyes some four years later in quatrain form, including the melody and chords for possible oral presentation![4]

In answer to those who accuse chemists of being illiterate with respect to the classical languages, I offer the example of a 1967 paper by T. K. Lim and M. A. Whitehead in presumably impeccable Latin. The opening paragraph of that paper begins:

> McWeeny erat primus, qui modo matricis densitatis (sic nominatur) [1-3], in quo continua iteratio eigenaequationis $FC_0 = SC_0E_0$ vitari potest, in quantum mechanica ratione usus est. In modo McWeeny matrix densitatis, quae mores systematis constituit sed tamen ad speciem singulorum orbitalium non advertit, declivissimi descensus modo iteratur.[5]

Among the many acronyms that have appeared in the literature are the following:

Ligand-**I**nduced **P**roton **S**hift (LIPS)
Lamb-**D**ip **S**pectroscopy
Proton **E**nhanced **N**uclear **I**nduction **S**pectroscopy

Fast, Accurate, Kinetic Energy (FAKE) Molecular Orbital Calculations

diaminomaleonitrile (DAMN). This report is literally sprinkled with DAMNs.

Calorific Recovery Anaerobic Process for converting cattle feedlot manure to useful products.[6]

Papers have been written by:

Sigma R. Alpha; Alexander, Baenziger, Carpenter, and Doyle (A, B, C, and D); Crystal Lin; Anne Bird Crow (in a study of "birdcage" alcohols. Her coauthor's middle name is "Thatcher."); Young and Auld; and Gray, Brown, and White.[7]

Authors have cited many nonscientific sources to establish or clarify a point. Perhaps the most frequently cited is the Bible, with references to Ecclesiastes, Joshua, Corinthians, and Matthew, among others. A nice example is provided by the Fiesers, who invoke Matthew to characterize the Saytzeff rule that in the dehydration of an alcohol, hydrogen is eliminated from the adjacent carbon atom poorest in hydrogen: "But from him that hath not, shall be taken away even that which he hath."[8] References to Shakespearean plays, nursery rhymes, fictional characters, and even such peculiarities as the citing of "Anonymous, C. and News, E." may be found.

Since many organic compounds have highly complicated names, it is hardly any wonder that it becomes irresistible to simplify the forbidding nomenclature. Names can be selected on the basis of the shape of a structural formula; for example, *manxane* corresponds to the symbol of the Island of Man, and *fenestrane* suggests the structure's resemblance to a window. Authors even take this pattern one step further with "broken-window compounds." Sometimes the author may wish to honor an admired mentor, as in, for example, *Louis Fieserone*.[9]

MANXANE   FENESTRANE

A name may be coined simply because it was irresistible, for example, *megaphone*, the name of a ketone isolated from the roots of *Aniba megaphylla*. Perhaps the most extreme case involves the investigators who identified the various anthracyclines isolated from their bohemic acid complex by names from an opera with a similar name.[10] Thus they reported mimimycin, alcindormycin, and musettamycin, among others.

One of my favorite examples of chemage appeared in the "Additions and Corrections" section of the *Journal of the American Chemical Society* in 1977. In a statement that an author's name had been misspelled, the word "mispelled" was itself misspelled. Although automobiles are routinely recalled, there is actually one case on record of a textbook being recalled for not dissimilar reasons.

```
91: 139857g NOTICE:   (The publisher has recalled all
copies of this book, which contains a serious and potentially
dangerous misprint in the description of an experiment).
A Modern Course of Organic Chemistry.  Gerrish, J. R.;
Whitfield, R. C. (Longman: Harlow, Engl.). 1977. 432 pp.
(Eng) £3.65.
```

From *Chemical Abstracts*, 91 (1979), 139857g.

This CAS citation is copyrighted by the American Chemical Society and is reprinted by permission. No further copying is allowed.

With the increasing use of data banks, it is no surprise that strange items can be found in them.

```
27060295    ID NO.- MGA27060295
   Chicken plucking as measure of tornado wind speed.
   Vonnegut, B.
   Atmos. Sci. Res. Ctr., State Univ. of N.Y., Albany
   Weatherwise, Boston, 28(5): 217, Oct. 1975. DAS, DLC
   CTRY OF PUBL:US
      Loomis'   experiment  to  determine  the  air  speed
   required to remove all the feathers of  a  chicken  to
   estimate  the  wind  speed  in  a  tornado  vortex  was
   conducted in a wind tunnel  instead  of  using  a  dead
   chicken as a ball shot out of a six pounder. In view of
   the fact that the force required to remove the feathers
   from  the  follicles  varies  over  a  wind  range in a
   complicated and unpredictable way and depends upon  the
   chicken's   condition   and   its   reaction   to   its
   environment,  the plucking is of doubtful  value  as  an
   index of tornado wind velocity.
      DESCRIPTORS:   Tornado  wind  velocities;  Wind speed
   estimation
```

From Lockheed Information Systems, *Meteorological & Geoastrophysical Abstracts*, File 29.

Reproduced by permission of the American Meteorological Society.

To end this report on a suitably silly note, I cite the following reference found in the widely used *Handbook of Chemistry and Physics: "Sea water,* see Water, sea."[11]

NOTES

1. References for answers: 1) G. Arnison [140 other names] . . . E. Zurfluh, "Elastic and Total Cross Section Measurement at the Cern Proton-Antiproton Collider," *Physics Letters*, 128B (1983), 336–42.   2) Jean P. Barrier *et al.*, "Ring Contraction of 2-Alkylidenecyclobutanols to Cyclopropyl Carbonyl Compounds," *Journal of the Chemical Society, Chemical Communications*, (1973), 103–104.   3) Tamar M. Gund *et al.*, "Computer Assisted Graph Theoretical Analysis of Complex Mechanistic Problems in Polycyclic Hydrocarbons: The Mechanism of Diamantane Formation from Various Pentacyclotetradecanes," *Journal of the American Chemical Society*, 97 (1975), 749.   5) R. Patel, J. Patel, V. S. Patel, and K. C. Patel, "Synthesis and Thermal Stability of Phenol-1,2-Dichloroethane Resins," *Die Angewandete Makromoleculare Chemie*, 90 (1980), 201–10.   6) W. J. Thomas and E. Crawley, "The Glair Glands and Oosteal of *Austro-potamobius pallipes* (Lereboullet)," *Experientia*, 31 (1975), 183–85, 534–37.   7) Philip C. Keller, David MacLean, and Riley Schaffer, "Final Assignment of B1(3) n.m.r. Resonance of Decaborane," *Chemical Communications*, (1965), 204.
2. A. T. Wilson and M. Calvin, "The Photosynthetic Cycle: CO_2 Dependent Transients," *Journal of the American Chemical Society*, 77 (1955), 5948–57.
3. J. F. Bunnett and Francis J. Kearley, Jr., "Comparative Mobility of Halogens in Reactions of Dihalobenzenes with Potassium Amide in Ammonia," *Journal of Organic Chemistry*, 36 (1971), 184.
4. Howard M. Shapiro, Milan Bier, and C. F. Zukoski, "Continuous Redox State Monitoring of Preserved Organs," paper presented at the 26th Annual Conference on Engineering in Medicine and Biology, Minneapolis, 1973; Howard M. Shapiro, "Fluorescent Dyes for Differential Counts by Flow Cytometry: Does Histochemistry Tell Us Much More than Cell Geometry?" *Journal of Histochemistry and Cytochemistry*, 25 (1977), 976–89.
5. T. K. Lin and M. A. Whitehead, "Modus Computandi Eigenvectores et Eigenaestimationes e Matrice Densitatis," *Theoretica Chimica Acta*, 7 (1967), 1.
6. Gus J. Palenik, Douglas A. Sullivan, and Datta V. Naik, "A Ligand-Induced Proton Shift (LIPS) in Two Cobaloxime Complexes . . . ," *Journal of the American Chemical Society*, 98 (1976), 1177–82; K. N. Rao and C. Weldon Mathews (eds.), *Molecular Spectroscopy: Modern Research* (New York, 1972), Chap. 1; A. Pines, M. G. Gibby, and J. S. Waugh, "Proton-Enhanced Nuclear Induction Spectroscopy: A Method for High Resolution NMR of Dilute Spins in Solids," *Journal of Chemical Physics*, 56 (1972), 1776–77; Frank E. Harris, Alfred Trautwein, and Joseph Delhalle, "FAKE Molecular-Orbital Calculations," *Chemical Physics Letters*, 72 (1980), 315–18; Richard F. Schuman, Willard E. Shearin, and Roger J. Tull, "Chemistry of HCN: 1. Formation and Reactions of *N*-(Aminomethylidene)diaminomaleonitrile, an HCN Pentamer and Prescursor to Adenine," *Journal of Organic Chemistry*, 44 (1979), 4532–36; *Chemical & Engineering News*, 58 (1980), 6.
7. Sigma R. Alpha, "A Novel Reaction between 3,5-Dinitroacetophenone-Acetone and Secondary Amines Yielding Naphthalenic Structures," *Journal of Organic Chemistry*, 38 (1973), 3136–39; R. J. Alexander *et al.*, "Metal-Olefin Compounds: I. The Preparation and Molecular Structure of Some Metal-

Olefin Compounds Containing Norbornadiene (Bicyclo[2.2.1]hepta-2,5-diene)," *Journal of the American Chemical Society*, 82 (1960), 535–38; Elliot N. Marvell and Crystal Lin, "The Cope Rearrangement of *cis*-2-Phenyl-vinylcyclopropane," *Tetrahedron Letters*, (1973), 2679–82; Anne Bird Crow and Weston Thatcher Borden, "Stereochemistry of the Base Catalyzed Ketonization of the Birdcage Alcohol," *Tetrahedron Letters*, (1967), 1967–70; Thomas E. Young and David S. Auld, "Biindolyls: II. 2-(3-Indolyl)-3*H*-pseudoindol-3-ones from the Condensation of Isatin α-Chlorides with Indoles," *Journal of Organic Chemistry*, 28 (1963), 418–21; W.S. Singleton, M.S. Gray, M.L. Brown, and J.L. White, "Chromatographically Homogeneous Lecithin from Egg Phospholipids," *Journal of the American Oil Chemists' Society*, 42 (1965), 53–56.

8. Louis F. Fieser and Mary Fieser, *Advanced Organic Chemistry* (New York, 1961), 140.

9. M. Doyle *et al.*, "Synthesis and Conformational Mobility of Bicyclo(3,3,3)-undecane (Manxane)," *Tetrahedron Letters*, (1970), 3619–22; Vlasios Georgian and Martin Saltzman, "Syntheses Directed Toward Saturated 'Flat' Carbon," *ibid.*, (1972), 4315–17; Xorge A. Dominguez *et al.*, "Louisfieserone, an Unusual Flavanone Derivative," *ibid.*, (1978), 429–32.

10. S. Morris Kupchan *et al.*, "New Cytotoxic Neolignans from *Aniba megaphylla* Mez.," *Journal of Organic Chemistry*, 43 (1978), 586–90; Terrence W. Doyle *et al.*, "Antitumor Agents from the Bohemic Acid Complex: 4. Structures of Rudolphomycin, Mimimycin, Collinemycin, and Alcindormycin," *Journal of the American Chemical Society*, 101 (1979), 7041–49 (see footnote 1 of the article).

11. Robert C. Weast (ed.), *Handbook of Chemistry and Physics* (59th ed.; Boca Raton, Fla., 1978), Sec. I, p. 44.

Notes on the Contributors

ALAN J. ROCKE is associate professor in the Department of History of Science at Case Western Reserve University. He received a Ph.D. degree in the history of science from the University of Wisconsin—Madison in 1975 and Deutscher Akademischer Austauschdienst fellowships in 1974–1975 and 1984 for research in Germany. In 1981 he was codirector of a Department of Energy summer workshop. He was awarded the 1982 Jack Youden Prize by the Chemical Division, American Society for Quality Control. He has written journal articles and a book, *Chemical Atomism in the Nineteenth Century: From Dalton to Cannizzaro* (Columbus, Ohio, 1984). Although his teaching and research interests extend broadly across the history of science and technology, he concentrates on nineteenth- and twentieth-century physical sciences; atomic and structure theories; environment, food, and energy technologies; and the proliferation of nuclear weapons.

JOHN H. WOTIZ is professor of chemistry at Southern Illinois University at Carbondale. Born in Czechoslovakia, he came to the United States in 1939 and became a citizen in 1944. He received a Ph.D. degree in 1948 from The Ohio State University. He has had faculty positions at the University of Pittsburgh and Marshall University and a position as research supervisor at Diamond Alkali Company. His research on the chemistry of acetylenic and allenic compounds led to a long list of publications and patents. About 1972 his research interest shifted to international chemistry education and the history of chemistry. He conducts, biennially, a summer history-of-chemistry tour through Europe. He was editor of the *Journal of Chemical Educa-*

tion column "The Story Behind the Story" and, in 1980 and 1981, was chairman of the Division of History of Chemistry of the American Chemical Society. He conceived and promoted the idea of the newly established Center for History of Chemistry, of whose advisory board he is a member. In 1982 he received the international Dexter Award for outstanding contributions to the history of chemistry.

SUSANNA RUDOFSKY, born in Prague, Czechoslovakia, of Austrian and Swiss-French parents, is fluent in German and French as well as English. She was educated at Southern Illinois University in chemistry and the history of chemistry and is currently employed in the Department of Molecular Genetics and Cell Biology at the University of Chicago.

STEPHEN FINNEY MASON is professor of chemistry at King's College and a fellow of the Royal Society in London. He was awarded a Ph.D. degree in 1947 and a doctor of science degree in 1967, both by Oxford University. He has been a tutor in chemistry and a demonstrator in the Museum for the History of Science at Oxford University; a postdoctoral research fellow at the Australian National University and at University College, London; a senior lecturer and reader at the University of Exeter; and a professor of chemistry at the University of East Anglia. He has published extensively on his research in chemical spectroscopy and in the history of science. He is the author of *Main Currents of Scientific Thought* (New York, 1956), and of a revised edition, titled *A History of the Sciences* (New York, 1962).

O. BERTRAND RAMSAY is professor and head of the Department of Chemistry at Eastern Michigan University, where he has been a member of the faculty since 1965. He received a Ph.D. degree in 1960 from the University of Pennsylvania and postdoctoral research experience at Georgia Institute of Technology and at Northwestern University. His laboratory research has concentrated on organic photochemistry. His research in the history of chemistry, stimulated while he was a National Science

Foundation faculty fellow at the University of Wisconsin—Madison in 1968 and 1969, has resulted in two books, *Van't Hoff–Le Bel Centennial* (Washington, D.C., 1975) and *Stereochemistry* (London, 1981). He spent a sabbatical leave at the University of Reading, England, in 1973 and 1974. He served as chairman, in 1973 and 1974, and program chairman, from 1975 through 1984, of the Division of History of Chemistry, American Chemical Society.

JOHN ALFRED HEITMANN received a Ph.D. degree in the history of science from Johns Hopkins University in 1983. Prior to that time, he was a chemist at two industrial laboratories in Louisiana. He has served as assistant director for programs at the Center for History of Chemistry, where he conducted several oral history interviews, and as lecturer at the University of Pennsylvania. In mid-1984 he became a faculty member in the Department of History at the University of Dayton. His book *Scientific and Technical Change in the Louisiana Sugar Industry, 1830–1910* will be published by the LSU Press in 1987.

LEON GORTLER is professor of chemistry at Brooklyn College of the City University of New York. He joined the faculty there in 1962 after receiving a Ph.D. degree in 1961 from Harvard University (with Paul Bartlett as his mentor) and postdoctoral experience at the University of California—Berkeley. He has engaged in research in free radical chemistry, with molecules with sterically hindered rotation, and since 1978, in the history of chemistry. Oral history has figured prominently in his research in the history of physical organic chemistry. He was chairman of the Division of History of Chemistry, American Chemical Society, in 1982–1983.

JAMES G. TRAYNHAM, professor of chemistry at Louisiana State University, was the initiator of the LSU Mardi Gras Symposium in Organic Chemistry series. He received a Ph.D. degree in 1950 from Northwestern University. Concurrently, he was a full-time faculty member at Denison University and a part-time postdoctoral research associate at The Ohio State University before going to Louisiana State University in 1953. His research has been concerned mainly with mechanisms of organic reactions but has been extended recently to include the history of organic chemistry. He spent two sabbatical leaves in Europe, at Eidgenössische Technische Hochschule in Zurich, Switzerland, supported by an American Chemical Society Petroleum Research Fund award, and at the Universität des Saarlandes, Saarbrucken, Germany, as a NATO senior fellow in science. For eight years, beginning in 1973, he was a part-time chemist, at best, while he served Louisiana State University as vice-chancellor for advanced studies and research and dean of the Graduate School. He is author of a supplementary text, *Organic Nomenclature: A Programmed Introduction* (Englewood Cliffs, N.J., 3rd ed., 1985) and is chairman-elect of the Division of History of Chemistry, American Chemical Society.

JACK H. STOCKER is professor of chemistry at the University of New Orleans, where he has been a faculty member since its founding in 1958. He received a Ph.D. degree from Tulane University in 1955. Prior to that time, he was a chemist in two industrial laboratories in Michigan. He engaged in postdoctoral research at Tulane University, the University of Heidelberg in Germany, and Oak Ridge National Laboratories, and he spent a sabbatical leave at the University of Lund in Sweden. Before joining the faculty of the University of New Orleans, he was a faculty member at the University of Southern Mississippi. He has engaged in experimental research on stereochemical and electrochemical aspects of the formation of pinacols. He has been active in the governance of the American Chemical Society at both the local and national levels, and he has been an American Chemical Society tour speaker for several years.

Index

Acree, Solomon F., 95, 102
Acronyms, chemical, 131–32
Alcohol, ethyl, 7
Alexander, Elliot R., 111
American Chemical Society, New Orleans section, 91
Amino acid series, L-, 46, 50
Ampère, André Marie, 36, 37, 38
Armstrong, Henry Edward: early promotion of nomenclature, 115; report on Geneva conference, 116–17, 118
Analyses of cane juices, 86–87
Anschütz, Richard, 32–33
Aschan, Ossian, 65–66
Audubon experiment station. *See* Louisiana Sugar Experiment Station
Avogadro, Amedeo, 36

Baeyer, Adolf: cyclohexane compounds, 56, 58–59; 1, 2-cyclohexanedicarboxylic acids, 61–62; strain theory, 58–59, 60; use of molecular models, 59–60; mentioned, 23
Barlow, W., 47
Bartlett, Paul D.: early career, 106; education, 104, 106; influenced by Kohler, 97; status of research group, 109; students, 111; mentioned, 113n17
Barton, Derek H. R.: cyclohexane, 55; knowledge of Hassel's papers, 56; letter to Hermans, 75; Nobel Prize, 54; personal perspective on *Experientia* paper, 75; mentioned, 74
Beans, Hal, 97, 105
Becnel, Lezin, 88
Beeson, Jasper L., 85, 87, 88
Beilstein, Friedrich Konrad, 118

Benzol Fest, 23, 24–25, 26
Berichte der Durstigen Chemischen Gesellschaft, 27
Berk, J., 74
Berson, Jerome, 109
Berthelot, Marcellin, 16, 17
Berthollet, Claude Louis de, 14
Berzelius, Jöns Jakob: dualistic theory, 14, 35–36; rational formulas, 6; view of atoms, 6; mentioned, 30
Biblical references, 132
Bierzeitung, 27
Bijvoet, J. M., 46
Bilicke, Constant, 66–67
Binding energy of enantiomers, 49–50
Biomolecular homochirality, 45–46, 50
Bird, Maurice, 84
Bischoff, Carl, 68
Blacet, Francis, 107
Böeseken, Jacob, 68–70, 72
Bogert, Marston, 97
Boltzman, Ludwig, 17
Bounty Law, 91
Branch, Gerald E. K., 103
Bravais, Auguste, 47
Bravais lattices, 47
Breslow, Ronald, 110
Brodie, Benjamin, 15, 16, 17
Brønsted, Johannes Nicolaus, 105
Brown, A. Crum, 118
Brown, Herbert C., 107, 110, 111
Brown, Weldon G., 107, 110
Browne, Charles A., 84
Bryn Mawr College, 96
Bunnett, Joseph F., 130

California Institute of Technology, 95, 96
Calvin, Melvin, 103

Cannizzaro, Stanislao, 37, 117
Chain of atoms: 14; parent, 117
Chemical Abstracts Service: influence on nomenclature rules, 120, 123; index names, 123–24
City and Guilds Institute, 86
Clausius, Rudolf, 15
Clerget's method, 89, 93*n*19
Coauthors, unusual, 132
College of William and Mary, 81
Columbia University, 95, 96
Conant, James Bryant, 96, 97, 104
Conforth, John W., 54
Cope, Arthur C., 96, 97, 109, 110
Cotton, A., 43
Couper, Archibald Scott, 30
Cram, Donald J., 110
Crawley, Josiah Thomas, 85
Crick, Francis, 55
Cross, William E., 84
Curtin, David Y., 109, 111
1,2-Cycloalkanediols, 70–72
Cyclodecanone, 74
Cyclohexane: calculations for conformers, 74; planar structure, 56, 58–59, 66–67; Sachse's proposal, 62–63

Dalton, John, 5, 6, 36
D- and L- series, 44–46
Davy, Sir Humphry, 6, 14
Descartes, René, 2
Delafosse, Gabriel, 39, 40
Derx, H. G., 70, 74
Diatomic hypothesis, 38
Dickinson, Roscoe, 67
DNA, 55
Doering, William von E., 109
Drude, P., 46
Dualistic theory, 35–36
Duhem, Pierre, 2–3
Duke University, 95
Dumas, Jean-Baptiste: chlorine substitution, 7; view on theories, 1–2; mentioned, 14, 21, 30, 37, 38

Einstein, Albert, 18
Elderfield, Robert, 107, 109
Eliel, Ernest L., 72
Enantiomorphs, 48
Epitomization analogy, 37, 39
Evans, William, 102

Falk, K. George, 97
Faraday, Michael, 41
Federov, E. S., 47
Fieser, Louis, 56, 96, 104
Fischer, Emil, 44–45, 46, 84
Fischer-Rosanoff convention, 44–46
Fisherman illustration, 130
Frankland, Edward, 21, 30, 37
Franklin, Rosalind, 55
Freund, Ida, 17, 18
Friedel, Charles, 117

Gaudin, Marc Antoine Augustin, 38
Geissman, Theodore A., 110
Geneva conference on nomenclature, 116–17, 118, 119–20
Gerhardt, Charles, 10, 11, 14, 18, 30
Gomberg, Moses, 95
Greene, Frederick D., 111
Guinness records for chemists, 129–30

Hammett, Louis P.: as administrator, 110; education, 104–105; influenced by Kohler, Nelson, 97; publications, 105, 106, 108; theory of acidity, 105; WWII work, 109; mentioned, 113*n*17
Hammett's acidity function, 105
Hantzsch, Arthur, 105
Harvard University, 79, 85, 86, 95, 96
Hassel, Odd, 54, 56
Haüy, René Just, 35, 37, 39, 40
Hendricks, Sterling B., 66–67
Hermans, Peter, 70, 74
Herschel, John W. F., 39
Hessel, J. F. C., 47
Hexabromocyclohexane, 66–67
Hiawatha-like benzyne paper, 130
Hinshelwood, Cyril Norman, 107
Hofmann, August Wilhelm von, 23, 115
Holness, N. J., 72
Horton, Horace Everett Lincoln, 85
Hückel, Walter, 65
Hughes, Edward D., 108

Illinois Institute of Technology, 110
Ingold, Christopher K., 66, 107
"Instructions to Abstractors," 115–16
International Chemical Congress, 1899, p. 116
International Union of Chemistry, 119

International Union of Pure and Applied Chemistry (IUPAC), 121–22, 123
Iowa State College, 95, 111

Jackson, Charles Loring, 96
Jacobs, Thomas, 110
Japp, Francis R., 42
Johns Hopkins University, 79, 85, 95
Johnson, Samuel W., 81
Jones, T. H., 84

Karlsruhe Congress (1860), 14, 37
Katz, Thomas, 110
Kekulé, Friedrich August: associations with other chemists, 11, 18; Benzol Fest speech, 24, 25, 30; biographical data, 21, 31; devotion to ontological theory, 13, 14; dreams, 12, 22, 24–29 *passim;* education, 11; family members, 26–27, 32, 33; family title, 32; influence of German nationalism, 30–31; letter to Baeyer, 23; letters to Schultz, 23, 24; publications about benzene structure, 22, 29; rational formulas, 13; tetravalent carbon, 11; mentioned, 39, 58
Kekulé-Baeyer models, 60, 62, 64
Kekule von Stradonitz, Stephen, 29, 32
Kharasch, Morris, 105–107
Kinetic theory, 15
Kohler, Elmer Peter, 95, 96, 97, 104, 105
Kolbe, Hermann, 12, 14, 16, 18
Kopp, Hermann, 29

La Mer, Victor K., 106
Landholt, Hans Heinrich, 90
Latin paper, 131
Laurent, August: chlorine substitution, 7; crystal-molecule analogies, 38; hexagon formulas, 25; relationship to Gerhardt, 11–12
Lavoisier, Antoine Laurent, 5, 35–36
Le Bel, Joseph Achille, 39, 43, 46, 47, 56–57
Levene, Phoebus Aaron, 106
Lewis, Gilbert Newton, 79, 103, 105
Liebig, Justus: benzoyl fragment, 114; disenchantment with theories, 8; equivalent weights, 9; formulation of ethyl alcohol, 7; mentioned, 30, 82, 91
Liége rules of nomenclature, 119–20
Lin, T. K., 131
Linstead, Reginald Patrick, 109
Loschmidt, Joseph, 31
Louisiana Sugar Chemists' Association, 88–89, 90–91
Louisiana Sugar Experiment Station, 81, 82, 87, 91
Louisiana Sugar Planters Association, 78, 80
Lucas, Howard J., 102, 103–104, 107
Lukach, Carl, 72

McCoy, Herbert, 102
Mach, Ernst, 18
Mallet, John William, 81
Mandelbaum, Maurice, 1, 2
Maxwell, James Clerk, 15, 17
Maxwell, Walter, 85, 86–88
Mayo, Frank, 107, 110, 113n17
Michael, Arthur, 95, 96
Mispelled, 133
Mitscherlich, Eilhard, 40
Mohr, Ernst Wilhelm, 64–65
Molecular models, 54, 55, 59–60
Monkey structure of benzene, 27

Names, chemical: benzenol, 121; benzoyl, 114; benzyl, 120; cycloalkanes, 117, 118; fenestrane, 132; formic acid, 114; Louis Fieserone, 132; manxane, 132; megaphone, 133; naphthyl, 120; parent chain, 117; phene, 121; phenol, 121; phenyl, 120; prefix *n-*, 117; puns, 128–29; radicles, 116; systematic endings, 115–17; trimethyl phosphate, 123
Naquet, Alfred, 16
Nef, John Ulrich, 95, 97, 102
Nelson, John M. "Pop," 97, 105, 106
Niemann, Carl, 107
Nobel prizes, 54
Noyce, Donald S., 109, 110
Noyes, Arthur A., 97, 103

Odling, William, 16, 17
Organic Reaction Mechanism Conference, first, 111, 113n17
Ostwald, Wilhelm, 17

Ouroboros, 25, 30
Oxford University, 107

Parity violation, 49
Pasteur, Louis, 31, 39, 40, 41–42
Patterson, Austin M., 120–21
Pauling, Linus, 55, 105, 107
Pharr, John N., 90
Physical organic chemists, 108, 109, 111
Planck, Max, 17
Polyhydroxy compounds, 68–70
Porter, Charles Walter, 103
Prelog, Vladimir, 54, 74–75
Price, Charles C., 97, 111, 113n17
Puns, chemical, 128–29

Quartz crystals, enantiomeric, 49

Radicles, 116
Randolph Macon College, 81
Remsen, Ira, 79, 96, 118
Richards, Theodore W., 79, 96, 97, 104
Roberts, John D., 110, 111
Robertson, G. Ross, 110
Rockefeller Institute, 106
Rosanoff, M. A., 45, 46
Ross, Bennett Battle, 89, 90
Rotational barrier, 67–68
Rules of nomenclature, 121–23
Russell, Colin A., 75
Ruzicka, Leopold, 66

S. C. H. Windler letter, 7–8
Sachse, Hermann, 62–63
Sachse's theory, 62–64, 66
Schlesinger, Herman, 107
Schoenflies, A., 47
Schultz, Gustav, 23
Schulze, Ernest, 86
Shapiro, Howard, 131
Smith, Alexander, 102
Sohncke, Leonhard, 47
Spoehr, Herman, 102
Stewart, T. D., 103
Stieglitz, Julius, 95, 97, 102, 105
Stork, Gilbert, 110
Streitwieser, Andrew, Jr., 109, 110
Stubbs, William Carter, 81–83
Sugarcane nodes, 85
Sugar chemists, 79–80, 82, 83, 88–90
Sugar houses, 78
Sugar industry, decline, 91

Sugar series, D-, 44–46, 50
Swain, C. Gardner, 111

"Talking blues" paper, 131
Tarbell, Stanley, 97
Teaching organic nomenclature, 124–25
Tetrahedral carbon, 46, 48
Tetratomic hypothesis, 38
Textbook recall, 133
Theories: caloric, 15, 36; crystal-molecule epitomization, 35; functions of, 2; polar dualistic, 35, 36; purpose, 1, 3
Theorists, classification of, 3, 4
Thermodynamics, 15
Thorpe, Jocelyn, 66
Throop College of Technology, 103
Tiemann, Ferdinand, 102, 116
Tishler, Max, 97
Tollens, Bernard, 84, 85
Transdiction, 2, 6, 12
Twain, Mark, 78

United States Department of Agriculture, 78, 80, 81
University of California, Berkeley, 103
University of California, Los Angeles, 95, 96, 106
University of Chicago, 79, 95, 96, 97
University of Illinois, 95, 111
University of Minnesota, 106
University of Notre Dame, 111
University of Virginia, 81
University of Wisconsin, 95, 111
Urry, Wilbert H., 110, 113n17

Valence, 11, 37
Van't Hoff, Jacobus H.: acidity of borate esters, 69; basic stereochemical ideas, 56–58; cis/trans isomers, 58; rotational barrier, 68; mentioned, 13, 16, 18, 39
Vitalism, 42, 44

Walling, Cheves, 110
Watson, James D., 55
Werner, Alfred, 39, 42, 105
Westheimer, Frank H., 97, 104, 107, 113n17
Wheland, George Willard, 104, 107, 113n17
Whitehead, M. A., 131

Whitmore, Frank C., 97
Wiberg, Kenneth B., 109
Will, Heinrich, 30
Williamson, Alexander: connection with Gerhardt, 10–11; criticism of Kolbe's formulas, 12; defense of atomic theory, 17; ether synthesis, 11, 18; influence on Kekulé, 13, 18; mentioned, 21, 30
Winstein, Saul, 72, 104, 108, 110, 113n17
Wipprecht, W., 84

Wislicenus, Johannes, 58
Wöhler, Friedrich, 7, 8, 81, 114
Woodward, Robert Burns, 108
Wurtz, Charles Adolphe, 30, 58

Yoder, Peter A., 84
Young, William G., 102, 104, 105, 106, 107

Zerban, Fritz, 84
Zurich Polytechnical Institute, 85, 86